**NONRESIDENT
TRAINING
COURSE**

SEPTEMBER 1998

Navy Electricity and Electronics Training Series

Module 14—Introduction to Microelectronics

NAVEDTRA 14186

Although the words "he," "him," and "his" are used sparingly in this course to enhance communication, they are not intended to be gender driven or to affront or discriminate against anyone.

PREFACE

By enrolling in this self-study course, you have demonstrated a desire to improve yourself and the Navy. Remember, however, this self-study course is only one part of the total Navy training program. Practical experience, schools, selected reading, and your desire to succeed are also necessary to successfully round out a fully meaningful training program.

COURSE OVERVIEW: To introduce the student to the subject of Microelectronics who needs such a background in accomplishing daily work and/or in preparing for further study.

THE COURSE: This self-study course is organized into subject matter areas, each containing learning objectives to help you determine what you should learn along with text and illustrations to help you understand the information. The subject matter reflects day-to-day requirements and experiences of personnel in the rating or skill area. It also reflects guidance provided by Enlisted Community Managers (ECMs) and other senior personnel, technical references, instructions, etc., and either the occupational or naval standards, which are listed in the Manual of Navy Enlisted Manpower Personnel Classifications and Occupational Standards, NAVPERS 18068.

THE QUESTIONS: The questions that appear in this course are designed to help you understand the material in the text.

VALUE: In completing this course, you will improve your military and professional knowledge. Importantly, it can also help you study for the Navy-wide advancement in rate examination. If you are studying and discover a reference in the text to another publication for further information, look it up.

1998 Edition Prepared by TDCS Paul H.Smith

Published by NAVAL EDUCATION AND TRAINING PROFESSIONAL DEVELOPMENT AND TECHNOLOGY CENTER

NAVSUP Logistics Tracking Number 0504-LP-026-8390

Sailor's Creed

"I am a United States Sailor.

I will support and defend the Constitution of the United States of America and I will obey the orders of those appointed over me.

I represent the fighting spirit of the Navy and those who have gone before me to defend freedom and democracy around the world.

I proudly serve my country's Navy combat team with honor, courage and commitment.

I am committed to excellence and the fair treatment of all."

TABLE OF CONTENTS

3. Miniature and Microminiature Repair Procedures 3-1

APPENDIX

I. Glossary AI-1

II. Reference List AII-1

INDEX INDEX-1

CREDITS

Many of the figures included in this edition of NEETS, Module 14, Introduction to Microelectronics, were provided by the 2M section of the Education and Training Division, Naval Air Rework Facility, Pensacola, Florida, and the Naval Undersea Warfare Engineering Center, Keyport, Washington. Their assistance is gratefully acknowledged.

Permission to use the trademark "PANAVISE" by Pana Vise Products, Inc., is gratefully acknowledged.

The illustrations indicated below were provided by the designated companies. Permission to use these illustrations is gratefully acknowledged:

SOURCE

FIGURE

Siliconix, Inc.

Pana Vise Products, Inc. (former company name: Harris Semiconductors)

1-33 1-34

NAVY ELECTRICITY AND ELECTRONICS TRAINING

SERIES

The Navy Electricity and Electronics Training Series (NEETS) was developed for use by personnel in many electrical- and electronic-related Navy ratings. Written by, and with the advice of, senior technicians in these ratings, this series provides beginners with fundamental electrical and electronic concepts through self-study. The presentation of this series is not oriented to any specific rating structure, but is divided into modules containing related information organized into traditional paths of instruction.

The series is designed to give small amounts of information that can be easily digested before advancing further into the more complex material. For a student just becoming acquainted with electricity or electronics, it is highly recommended that the modules be studied in their suggested sequence. While there is a listing of NEETS by module title, the following brief descriptions give a quick overview of how the individual modules flow together.

Module 1, Introduction to Matter, Energy, and Direct Current, introduces the course with a short history of electricity and electronics and proceeds into the characteristics of matter, energy, and direct current (dc). It also describes some of the general safety precautions and first-aid procedures that should be common knowledge for a person working in the field of electricity. Related safety hints are located throughout the rest of the series, as well.

Module 2, Introduction to Alternating Current and Transformers, is an introduction to alternating current (ac) and transformers, including basic ac theory and fundamentals of electromagnetism, inductance, capacitance, impedance, and transformers.

Module 3, Introduction to Circuit Protection, Control, and Measurement, encompasses circuit breakers, fuses, and current limiters used in circuit protection, as well as the theory and use of meters as electrical measuring devices.

Module 4, Introduction to Electrical Conductors, Wiring Techniques, and Schematic Reading, presents conductor usage, insulation used as wire covering, splicing, termination of wiring, soldering, and reading electrical wiring diagrams.

Module 5, Introduction to Generators and Motors, is an introduction to generators and motors, and covers the uses of ac and dc generators and motors in the conversion of electrical and mechanical energies.

Module 6, Introduction to Electronic Emission, Tubes, and Power Supplies, ties the first five modules together in an introduction to vacuum tubes and vacuum-tube power supplies.

Module 7, Introduction to Solid-State Devices and Power Supplies, is similar to module 6, but it is in reference to solid-state devices.

Module 8, Introduction to Amplifiers, covers amplifiers.

Module 9, Introduction to Wave-Generation and Wave-Shaping Circuits, discusses wave generation and wave-shaping circuits.

Module 10, Introduction to Wave Propagation, Transmission Lines, and Antennas, presents the characteristics of wave propagation, transmission lines, and antennas.

Module 11, Microwave Principles, explains microwave oscillators, amplifiers, and waveguides. Module 12, Modulation Principles, discusses the principles of modulation.

Module 13, Introduction to Number Systems and Logic Circuits, presents the fundamental concepts of number systems, Boolean algebra, and logic circuits, all of which pertain to digital computers.

Module 14, Introduction to Microelectronics, covers microelectronics technology and miniature and microminiature circuit repair.

Module 15, Principles of Synchros, Servos, and Gyros, provides the basic principles, operations, functions, and applications of synchro, servo, and gyro mechanisms.

Module 16, Introduction to Test Equipment, is an introduction to some of the more commonly used test equipments and their applications.

Module 17, Radio-Frequency Communications Principles, presents the fundamentals of a radio-frequency communications system.

Module 18, Radar Principles, covers the fundamentals of a radar system.

Module 19, The Technician's Handbook, is a handy reference of commonly used general information, such as electrical and electronic formulas, color coding, and naval supply system data.

Module 20, Master Glossary, is the glossary of terms for the series.

Module 21, Test Methods and Practices, describes basic test methods and practices.

Module 22, Introduction to Digital Computers, is an introduction to digital computers.

Module 23, Magnetic Recording, is an introduction to the use and maintenance of magnetic recorders and the concepts of recording on magnetic tape and disks.

Module 24, Introduction to Fiber Optics, is an introduction to fiber optics.

Embedded questions are inserted throughout each module, except for modules 19 and 20, which are reference books. If you have any difficulty in answering any of the questions, restudy the applicable section.

Although an attempt has been made to use simple language, various technical words and phrases have necessarily been included. Specific terms are defined in Module 20, Master Glossary.

Considerable emphasis has been placed on illustrations to provide a maximum amount of information. In some instances, a knowledge of basic algebra may be required.

Assignments are provided for each module, with the exceptions of Module 19, The Technician's Handbook; and Module 20, Master Glossary. Course descriptions and ordering information are in NAVEDTRA 12061, Catalog of Nonresident Training Courses.

Throughout the text of this course and while using technical manuals associated with the equipment you will be working on, you will find the below notations at the end of some paragraphs. The notations are used to emphasize that safety hazards exist and care must be taken or observed.

WARNING

AN OPERATING PROCEDURE, PRACTICE, OR CONDITION, ETC., WHICH MAY RESULT IN INJURY OR DEATH IF NOT CAREFULLY OBSERVED OR FOLLOWED.

CAUTION

AN OPERATING PROCEDURE, PRACTICE, OR CONDITION, ETC., WHICH MAY RESULT IN DAMAGE TO EQUIPMENT IF NOT CAREFULLY OBSERVED OR FOLLOWED.

NOTE

An operating procedure, practice, or condition, etc., which is essential to emphasize.

INSTRUCTIONS FOR TAKING THE COURSE

ASSIGNMENTS

The text pages that you are to study are listed at the beginning of each assignment. Study these pages carefully before attempting to answer the questions. Pay close attention to tables and illustrations and read the learning objectives. The learning objectives state what you should be able to do after studying the material. Answering the questions correctly helps you accomplish the objectives.

SELECTING YOUR ANSWERS

Read each question carefully, then select the BEST answer. You may refer freely to the text. The answers must be the result of your own work and decisions. You are prohibited from referring to or copying the answers of others and from giving answers to anyone else taking the course.

SUBMITTING YOUR ASSIGNMENTS

To have your assignments graded, you must be enrolled in the course with the Nonresident Training Course Administration Branch at the Naval Education and Training Professional Development and Technology Center (NETPDTC). Following enrollment, there are two ways of having your assignments graded: (1) use the Internet to submit your assignments as you complete them, or (2) send all the assignments at one time by mail to NETPDTC.

Grading on the Internet: Advantages to Internet grading are:

• you may submit your answers as soon as you complete an assignment, and

• you get your results faster; usually by the next working day (approximately 24 hours).

In addition to receiving grade results for each assignment, you will receive course completion confirmation once you have completed all the

assignments. To submit your assignment answers via the Internet, go to:

http ://courses.cnet.na vy.mil

Grading by Mail: When you submit answer sheets by mail, send all of your assignments at one time. Do NOT submit individual answer sheets for grading. Mail all of your assignments in an envelope, which you either provide yourself or obtain from your nearest Educational Services Officer (ESO). Submit answer sheets to:

COMMANDING OFFICER NETPDTC N331 6490 SAUFLEY FIELD ROAD PENSACOLA FL 32559-5000

Answer Sheets: All courses include one "scannable" answer sheet for each assignment. These answer sheets are preprinted with your SSN, name, assignment number, and course number. Explanations for completing the answer sheets are on the answer sheet.

Do not use answer sheet reproductions: Use

only the original answer sheets that we provide—reproductions will not work with our scanning equipment and cannot be processed.

Follow the instructions for marking your answers on the answer sheet. Be sure that blocks 1, 2, and 3 are filled in correctly. This information is necessary for your course to be properly processed and for you to receive credit for your work.

COMPLETION TIME

Courses must be completed within 12 months from the date of enrollment. This includes time required to resubmit failed assignments.

PASS/FAIL ASSIGNMENT PROCEDURES

If your overall course score is 3.2 or higher, you will pass the course and will not be required to resubmit assignments. Once your assignments have been graded you will receive course completion confirmation.

If you receive less than a 3.2 on any assignment and your overall course score is below 3.2, you will be given the opportunity to resubmit failed assignments. You may resubmit failed assignments only once. Internet students will receive notification when they have failed an assignment—they may then resubmit failed assignments on the web site. Internet students may view and print results for failed assignments from the web site. Students who submit by mail will receive a failing result letter and a new answer sheet for resubmission of each failed assignment.

COMPLETION CONFIRMATION

After successfully completing this course, you will receive a letter of completion.

ERRATA

Errata are used to correct minor errors or delete obsolete information in a course. Errata may also be used to provide instructions to the student. If a course has an errata, it will be included as the first page(s) after the front cover. Errata for all courses can be accessed and viewed/downloaded at:

http://www.advancement.cnet.navy.mil

STUDENT FEEDBACK QUESTIONS

We value your suggestions, questions, and criticisms on our courses. If you would like to communicate with us regarding this course, we encourage you, if possible, to use e-mail. If you write or fax, please use a copy of the Student Comment form that follows this page.

For subject matter questions:

E-mail: n315.products@cnet.navy.mil Phone: Comm: (850) 452-1001, ext. 1728 DSN: 922-1001, ext. 1728 FAX: (850)452-1370 (Do not fax answer sheets.) Address: COMMANDING OFFICER NETPDTC N315 6490 SAUFLEY FIELD ROAD PENSACOLA FL 32509-5237

For enrollment, shipping, grading, or completion letter questions

E-mail: fleetservices @ cnet. navy. mil

Phone: Toll Free: 877-264-8583

Comm: (850)452-1511/1181/1859 DSN: 922-1511/1181/1859 FAX: (850)452-1370 (Do not fax answer sheets.)

Address: COMMANDING OFFICER NETPDTC N331 6490 SAUFLEY FIELD ROAD PENSACOLA FL 32559-5000

NAVAL RESERVE RETIREMENT CREDIT

If you are a member of the Naval Reserve, you will receive retirement points if you are authorized to receive them under current directives governing retirement of Naval Reserve personnel. For Naval Reserve retirement, this course is evaluated at 4 points. (Refer to Administrative Procedures for Naval Reservists on Inactive Duty, BUPERSINST 1001.39, for more information about retirement points.)

THIS PAGE LEFT BLANK INTENTIONALLY.

Student Comments

NEETS Module 14 Course Title: Introduction to Microelectronics

NAVEDTRA: 14186 Date:

We need some information about you :

Rate/Rank and Name: SSN: Command/Unit

Street Address: City: State/FPO: Zip

Your comments, suggestions, etc.:

Privacy Act Statement: Under authority of Title 5, USC 301, information regarding your military status is requested in processing your comments and in preparing a reply. This information will not be divulged without written authorization to anyone other than those within POD for official use in determining performance.

NETPDTC 1550/41 (Rev 4-00)

CHAPTER 1

MICROELECTRONICS

LEARNING OBJECTIVES

Learning objectives are stated at the beginning of each topic. These learning objectives serve as a preview of the information you are expected to learn in the topic. The comprehensive check questions are based on the objectives. By successfully completing the OCC-ECC, you indicate that you have met the objectives and have learned the information. The learning objectives are listed below.

Upon completion of this topic, you will be able to:

1. Outline the progress made in the history of microelectronics.

2. Describe the evolution of microelectronics from point-to-point wiring through

high element density state-of-the-art microelectronics.

     3. List the advantages and disadvantages of point-to-point wiring and high element density state-of-the-art microelectronics.

     4. Identify printed circuit boards, diodes, transistors, and the various types of integrated circuits. Describe the fabrication techniques of these components.

     5. Define the terminology used in microelectronic technology including the following terms used by the Naval Systems Commands:

     6. Describe typical packaging levels presently used for microelectronic systems.

     7. Describe typical interconnections used in microelectronic systems.

     8. Describe environmental considerations for microelectronics.

INTRODUCTION

In NEETS, Module 6, Introduction to Electronic Emission, Tubes, and Power Supplies, you learned that Thomas Edison's discovery of thermionic emission opened the door to electronic technology. Progress was slow in the beginning, but each year brought new and more amazing discoveries. The development of vacuum tubes soon led to the simple radio. Then came more complex systems of communications. Modern systems now allow us to communicate with other parts of the world via satellite. Data is now collected from space by probes without the presence of man because of microelectronic technology.

Sophisticated control systems allow us to operate equipment by remote control in hazardous situations, such as the handling of radioactive materials. We can remotely pilot aircraft from takeoff to landing. We can make course corrections to spacecraft millions of miles from Earth. Space flight, computers, and even video games would not be possible except for the advances made in microelectronics.

The most significant step in modern electronics was the development of the transistor by Bell Laboratories in 1948. This development was to solid-state electronics what the Edison Effect was to the vacuum tube. The solid-state diode and the transistor opened the door to microelectronics.

MICROELECTRONICS is defined as that area of technology associated with and applied to the realization of electronic systems made of extremely small electronic parts or elements. As discussed in topic 2 of NEETS, Module 7, Introduction to Solid-State Devices and Power Supplies, the term microelectronics is normally associated with integrated circuits (IC). Microelectronics is often thought to include only integrated circuits. However, many other types of circuits also fall into the microelectronics category. These will be discussed in greater detail under solid-state devices later in this topic.

During World War II, the need to reduce the size, weight, and power of military electronic systems became important because of the increased use of these systems. As systems became more complex, their size, weight, and power requirements rapidly increased. The increases finally reached a point that was unacceptable, especially in aircraft and for infantry personnel who carried equipment in combat. These unacceptable factors were the driving force in the development of smaller, lighter, and more efficient electronic circuit components. Such requirements continue to be important factors in the development of new systems, both for military and commercial markets. Military electronic systems, for example, continue to become more highly developed as their capability, reliability, and maintainability is increased. Progress in the development of military systems and steady advances in technology point to an ever-increasing need for increased technical knowledge of microelectronics in your Navy job.

Q1. What problems were evident about military electronic systems during World War II?

Q2. What discovery opened the door to solid-state electronics?

Q3. What is microelectronics?

EVOLUTION OF MICROELECTRONICS

The earliest electronic circuits were fairly simple. They were composed of a few tubes, transformers, resistors, capacitors, and wiring. As more was learned by designers, they began to increase both the size and complexity of circuits. Component limitations were soon identified as this technology developed.

VACUUM-TUBE EQUIPMENT

Vacuum tubes were found to have several built-in problems. Although the tubes were lightweight, associated components and chassis were quite heavy. It was not uncommon for such chassis to weigh 40 to 50 pounds. In addition, the tubes generated a lot of heat, required a warm-up time from 1 to 2 minutes, and required hefty power supply voltages of 300 volts dc and more.

No two tubes of the same type were exactly alike in output characteristics. Therefore, designers were required to produce circuits that could work with any tube of a particular type. This meant that additional components were often required to tune the circuit to the output characteristics required for the tube used.

Figure 1-1 shows a typical vacuum-tube chassis. The actual size of the transformer is approximately 4x4x3 inches. Capacitors are approximately 1 x 3 inches. The components in the figure are very large when compared to modern microelectronics.

Figure 1-1.—Typical vacuum tube circuit.

A circuit could be designed either as a complete system or as a functional part of a larger system. In complex systems, such as radar, many separate circuits were needed to accomplish the desired tasks. Multiple-function tubes, such as dual diodes, dual triodes, tetrodes, and others helped considerably to reduce the size of circuits. However, weight, heat, and power consumption continued to be problems that plagued designers.

Another major problem with vacuum-tube circuits was the method of wiring components referred to as POINT-TO-POINT WIRING. Figure 1 -2 is an excellent example of point-to-point wiring. Not only did this wiring look like a rat's nest, but it often caused unwanted interactions between components. For example, it was not at all unusual to have inductive or capacitive effects between wires. Also, point-to-point wiring posed a safety hazard when troubleshooting was performed on energized circuits because of exposed wiring and test points. Point-to-point wiring was usually repaired with general purpose test equipment and common hand tools.

Figure 1-2.—Point-to-point wiring.

Vacuum-tube circuits proved to be reliable under many conditions. Still, the drawbacks of large size, heavy weight, and significant power consumption made them undesirable in most situations. For example, computer systems using tubes were extremely large and difficult to maintain. ENIAC, a completely electronic computer built in 1945, contained 18,000 tubes. It often required a full day just to locate and replace faulty tubes.

In some applications, we are still limited to vacuum tubes. Cathode-ray tubes used in radar, television, and oscilloscopes do not, as yet, have solid-state counterparts.

One concept that eased the technician's job was that of MODULAR PACKAGING. Instead of building a system on one large chassis, it was built of MODULES or blocks. Each module performed a necessary function of the system. Modules could easily be removed and replaced during troubleshooting and repair. For instance, a faulty power supply could be exchanged with a good one to keep the system operational. The faulty unit could then be repaired while out of the system. This is an example of how the module concept improved the efficiency of electronic systems. Even with these advantages, vacuum tube modules still had faults. Tubes and point-to-point wiring were still used and excessive size, weight, and power consumption remained as problems to be overcome.

Vacuum tubes were the basis for electronic technology for many years and some are still with us. Still, emphasis in vacuum-tube technology is rapidly becoming a thing of the past. The emphasis of technology is in the field of microelectronics.

Q4. What discovery proved to be the foundation for the development of the vacuum tube?

Q5. Name the components which greatly increase the weight of vacuum-tube circuitry.

Q6. What are the disadvantages of point-to-point wiring?

Q7. What is a major advantage of modular construction?

Q8. When designing vacuum-tube circuits, what characteristics of tubes must be taken into consideration?

SOLID-STATE DEVICES

Now would be a good time for you to review the first few pages of NEETS, Module 7, Introduction to Solid-State Devices and Power Supplies, as a refresher for solid-state devices.

The transition from vacuum tubes to solid-state devices took place rapidly. As new types of transistors and diodes were created, they were adapted to circuits. The reductions in size, weight, and power use were impressive. Circuits that earlier weighed as much as 50 pounds were reduced in weight to just a few ounces by replacing bulky components with the much lighter solid-state devices.

The earliest solid-state circuits still relied on point-to-point wiring which caused many of the disadvantages mentioned earlier. A metal chassis, similar to the type used with tubes, was required to provide physical support for the components. The solid-state chassis was still considerably smaller and lighter than the older, tube chassis. Still greater improvements in component mounting methods were yet to come.

One of the most significant developments in circuit packaging has been the PRINTED CIRCUIT BOARD (pcb), as shown in figure 1-3. The pcb is usually an epoxy board on which the circuit leads have been added by the PHOTOETCHING process. This process is similar to photography in that copper-clad boards are exposed to controlled light in the desired circuit pattern and then etched to remove the unwanted copper. This process leaves copper strips (LANDS) that are used to connect the components. In general, printed circuit boards eliminate both the heavy, metal chassis and the point-to-point wiring.

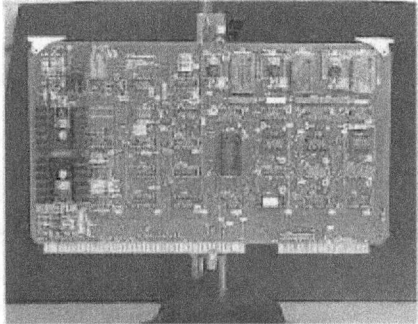

Figure 1-3.—Printed circuit board (pcb).

Although printed circuit boards represent a major improvement over tube technology, they are not without fault. For example, the number of components on each board is limited by the sizes and shapes of components. Also, while vacuum tubes are easily removed for testing or replacement, pcb components are soldered into place and are not as easily removed.

Normally, each pcb contains a single circuit or a subassembly of a system. All printed circuit boards within the system are routinely interconnected through CABLING HARNESSES (groups of wiring or ribbons of wiring). You may be confronted with problems in faulty harness connections that affect system reliability. Such problems are often caused by wiring errors, because of the large numbers of wires in a harness, and by damage to those wires and connectors.

Another mounting form that has been used to increase the number of components in a given space is the CORDWOOD MODULE, shown in figure 1-4. You can see that the components are placed perpendicular to the end plates. The components are packed very closely together, appearing to be stacked like cordwood for a fireplace. The end plates are usually small printed circuit boards, but may be insulators and solid wire, as shown in the figure. Cordwood modules may or may not be ENCAPSULATED (totally imbedded in solid material) but in either case they are difficult to repair.

Figure 1-4.—Cordwood module.

Q9. List the major advantages of printed circuit boards. Q10. What is the major disadvantage of printed circuit boards?

Qll. The ability to place more components in a given space is an advantage of the

.

INTEGRATED CIRCUITS

Many advertisements for electronic equipment refer to integrated circuits or solid-state technology. You know the meaning of the term solid-state, but what is an INTEGRATED CIRCUIT? The accepted Navy definition for an integrated circuit is that it consists of elements inseparably associated and formed on or within a single SUBSTRATE (mounting surface). In other words, the circuit components and all interconnections are formed as a unit. You will be concerned with three types of integrated circuits: MONOLITHIC, FILM, and HYBRID.

MONOLITHIC INTEGRATED CIRCUITS are those that are formed completely within a semiconductor substrate. These integrated circuits are commonly referred to as SILICON CHIPS.

FILM INTEGRATED CIRCUITS are broken down into two categories, THIN FILM and THICK FILM. Film components are made of either conductive or nonconductive material that is deposited in desired patterns on a ceramic or glass substrate. Film can only be used as passive circuit components, such as resistors and capacitors. Transistors and/or diodes are added to the substrate to complete the circuit. Differences in thin and thick film will be discussed later in this topic.

HYBRID INTEGRATED CIRCUITS combine two or more integrated circuit types or combine one or more integrated circuit types and DISCRETE (separate) components. Figure 1-5 is an example of a hybrid integrated circuit consisting of silicon chips and film circuitry. The two small squares are chips and the irregularly shaped gray areas are film components.

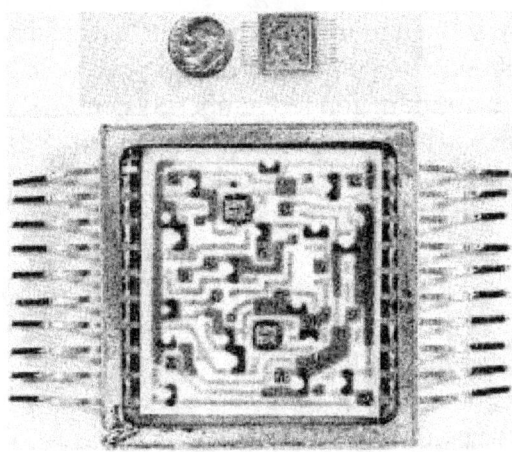

Figure 1-5.—Hybrid integrated circuit.

STATE-OF-THE-ART MICROELECTRONICS.

Microelectronic technology today includes thin film, thick film, hybrid, and integrated circuits and combinations of these. Such circuits are applied in DIGITAL, SWITCHING, and LINEAR (analog) circuits. Because of the current trend of producing a number of circuits on a single chip, you may look for further increases in the packaging density of electronic circuits. At the same time you may expect a reduction in the size, weight, and number of connections in individual systems. Improvements in reliability and system capability are also to be expected.

Thus, even as existing capabilities are being improved, new areas of microelectronic use are being explored. To predict where all this use of technology will lead is impossible. However, as the demand for increasingly effective electronic systems continues, improvements will continue to be made in state-of-the-art microelectronics to meet the demands.

LARGE-SCALE INTEGRATION (lsi) and VERY LARGE-SCALE INTEGRATION (vlsi) are the results of improvements in microelectronics production technology. Figure 1 -6 is representative of lsi. As shown in the figure, the entire SUBSTRATE WAFER (slice of semiconductor or insulator material) is

used instead of one that has been separated into individual circuits. In lsi and vlsi, a variety of circuits can be implanted on a wafer resulting in further size and weight reduction. ICs in modern computers, such as home computers, may contain the entire memory and processing circuits on a single substrate.

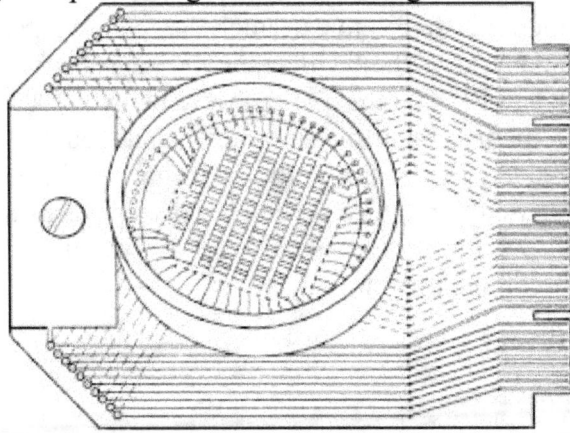

Figure 1-6.—Large-scale integration device (lsi).

Large-scale integration is generally applied to integrated circuits consisting of from 1,000 to 2,000 logic gates or from 1,000 to 64,000 bits of memory. A logic gate, as you should recall from NEETS, Module 13, Introduction to Number Systems, Boolean Algebra, and Logic Circuits, is an electronic switching network consisting of combinations of transistors, diodes, and resistors. Very large-scale integration is used in integrated circuits containing over 2,000 logic gates or greater than 64,000 bits of memory.

Q12. Define integrated circuit.

Q13. What are the three major types of integrated circuits?

Q14. How do monolithic ICs differ from film ICs?

QI5. What is a hybrid IC?

QI6. How many logic gates could be contained in lsi?

FABRICATION OF MICROELECTRONIC DEVICES

The purpose of this section is to give you a simplified overview of the manufacture of microelectronic devices. The process is far more complex than will be described here. Still, you will be able to see that microelectronics is not magic, but a highly developed technology.

Development of a microelectronic device begins with a demand from industry or as the result of research. A device that is needed by industry may be a simple diode network or a complex circuit consisting of thousands of components. No matter how complex the device, the basic steps of production are similar. Each type of device requires circuit design, component arrangement, preparation of a substrate, and the depositing of proper materials on the substrate.

The first consideration in the development of a new device is to determine what the device is to accomplish. Once this has been decided, engineers can design the device. During the design phase, the engineers will determine the numbers and types of components and the interconnections, needed to complete the planned circuit.

COMPONENT ARRANGEMENT

Planning the component arrangement for a microelectronic device is a very critical phase of production. Care must be taken to ensure the most efficient use of space available. With simple devices, this can be accomplished by hand. In other words, the engineers can prepare drawings of component placement. However, a computer is used to prepare the layout for complex devices. The computer is able to store the characteristics of thousands of components and can provide a printout of the most efficient component placement. Component placement is then transferred to extremely large drawings. During this step, care is taken to maintain the patterns as they will appear on the substrate. Figure 1-7 shows a fairly simple IC MASK PATTERN. If this pattern were being prepared for production, it would be drawn several hundred times the size shown and then photographed. The photo would then be reduced in size until it was the actual desired size. At that time, the pattern would be used to produce several hundred patterns that would be used on one substrate. Figure 1-8 illustrates how the patterns would be distributed to act as a WAFER MASK for manufacturing.

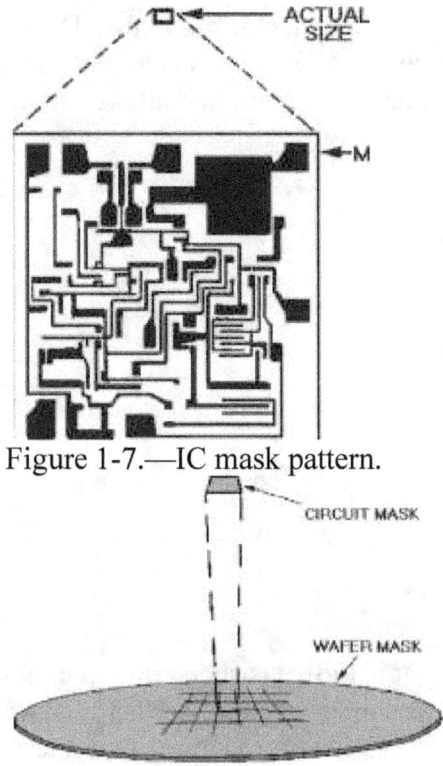

Figure 1-7.—IC mask pattern.

Figure 1-8.—Wafer mask distribution.

A wafer mask is a device used to deposit materials on a substrate. It allows material to be deposited in certain areas, but not in others. By changing the pattern of the mask, we can change the component arrangement of the circuit. Several different masks may be used to produce a simple microelectronic device. When used in proper sequence, conductor, semiconductor, or insulator materials may be applied to the substrate to form transistors, resistors, capacitors, and interconnecting leads.

SUBSTRATE PRODUCTION

As was mentioned earlier in this topic, microelectronic devices are produced on a substrate. This substrate will be of either insulator or semiconductor material, depending on the type of device. Film and hybrid ICs are normally constructed on a glass or ceramic substrate. Ceramic is usually the preferred material because of its durability.

Substrates used in monolithic ICs are of semiconductor material, usually silicon. In this type of IC, the substrate can be an active part of the IC. Glass or ceramic substrates are used only to provide support for the components.

Semiconductor substrates are produced by ARTIFICIALLY GROWING cylindrical CRYSTALS of pure silicon or germanium. Crystals are "grown" on a SEED CRYSTAL from molten material by slowly lifting and cooling the material repeatedly. This process takes place under rigidly controlled atmospheric and temperature conditions.

Figure 1 -9 shows a typical CRYSTAL FURNACE. The seed crystal is lowered until it comes in contact with the molten material-silicon in this case. It is then rotated and raised very slowly. The seed crystal is at a lower temperature than the molten material. When the molten material is in contact with the seed, it solidifies around the seed as the seed is lifted. This process continues until the grown crystal is of the desired length. A typical crystal is about 2 inches in diameter and 10 to 12 inches long. Larger diameter crystals can be grown to meet the needs of the industry. The purity of the material is strictly controlled to maintain specific semiconductor properties. Depending on the need, n or p impurities are added to produce the desired characteristics. Several other methods of growing crystals exist, but the basic concept of crystal production is the same.

RAISE
I

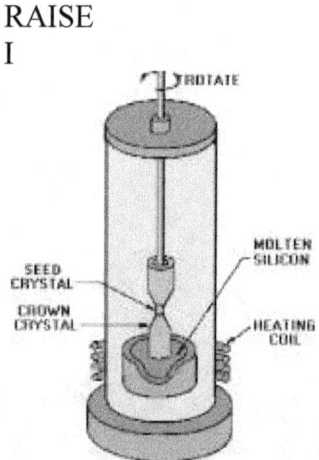

Figure 1-9.—Crystal furnace.

The cylinder of semiconductor material that is grown is sliced into thicknesses of .010 to .020 inch in the first step of preparation, as shown in figure 1-10. These wafers are ground and polished to remove any irregularities and to provide the smoothest surface possible. Although both sides are polished, only the side that will receive the components must have a perfect finish.

)
j LENGTH
10" T012"

Figure 1-10.—Silicon crystal and wafers.

Ql 7. What are the basic steps in manufacturing an IC?

Q18. Computer-aided layout is used to prepare devices.

Q19. What purpose do masks serve?

Q20. What type of substrates are used for film and hybrid ICs?

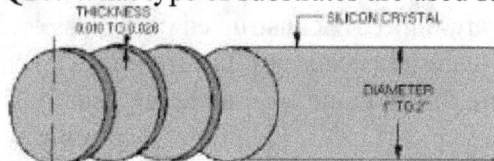

Q21. Describe the preparation of a silicon substrate.

FABRICATION OF IC DEVICES

Fabrication of monolithic ICs is the most complex aspect of microelectronic devices we will discuss. Therefore, in this introductory module, we will try to simplify this process as much as possible. Even though the discussion is very basic, the intent is still to increase your appreciation of the progress in microelectronics. You should, as a result of this discussion, come to realize that advances in manufacturing techniques are so rapid that staying abreast of them is extremely difficult.

Monolithic Fabrication.

Two types of monolithic fabrication will be discussed. These are the DIFFUSION METHOD and the EPITAXIAL METHOD.

DIFFUSION METHOD.—The DIFFUSION process begins with the highly polished silicon wafer being placed in an oven (figure 1-11). The oven contains a concentration impurity made up of impurity atoms which yield the desired electrical characteristics. The concentration of impurity atoms is diffused into the wafer and is controlled by controlling the temperature of the oven and the time that the silicon wafer is allowed to remain in the oven. This is called DOPING. When the wafer has been uniformly doped, the fabrication of semiconductor devices may begin. Several hundred circuits are produced simultaneously on the wafer.

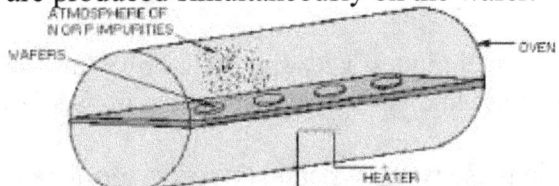

Figure 1-11.—Wafers in a diffusion oven.

The steps in the fabrication process described here, and illustrated in figure 1-12, would produce an npn, planar-diffused transistor. But, with slight variations, the technique may also be applied to the production of a complete circuit, including diodes, resistors, and capacitors. The steps are performed in the following order:

OXIDE COATING
STARTING MATERIAL N-TVPE

EMITTER DIFFUSION AN OXIDE FORMATION

OXIDE REMOVED FOR BASE DIFFUSION
gMITTEffe N
BASE-P
COLLECTOR-N
OXIDE REMOVED FOR EMITTER AND BASE CONTACTS
P-TVPE BORON DIFFUSION
N
BASE DIFFUSIONS AND
OXIDE FORMATION DURING BASE DIFFUSION

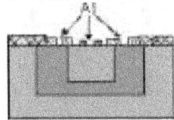

ALUMINUM f AL1 DEPOSITION AND ALLOYING TO FORM LEADS
OXIDE REMOVED FOR EMITTER DIFFUSION

Figure 1-12.—Planar-diffused transistor.

1. An oxide coating is thermally grown over the n-type silicon starting material.

2. By means of the photolithographic process, a window is opened through the oxide layer. This is done through the use of masks, as discussed earlier.

3. The base of the transistor is formed by placing the wafer in a diffusion furnace containing a p-type impurity, such as boron. By controlling the temperature of the oven and the length of time that the wafer is in the oven, you can control the amount of boron

diffused through the window (the boron will actually spread slightly beyond the window opening). A new oxide layer is then allowed to form over the area exposed by the window.

4. A new window, using a different mask much smaller than the first, is opened through the new oxide layer.

5. An n-type impurity, such as phosphorous, is diffused through the new window to form the emitter portion of the transistor. Again, the diffused material will spread slightly beyond the window opening. Still another oxide layer is then allowed to form over the window.

6. By means of precision-masking techniques, very small windows (about 0.005 inch in diameter) are opened in both the base and emitter regions of the transistor to provide access for electrical currents.

7. Aluminum is then deposited in these windows and alloyed to form the leads of the transistor or the IC.

(Note that the pn junctions are covered throughout the fabrication process by an oxide layer that prevents contamination.)

EPITAXIAL METHOD.—The EPITAXIAL process involves depositing a very thin layer of silicon to form a uniformly doped crystalline region (epitaxial layer) on the substrate. Components are produced by diffusing appropriate materials into the epitaxial layer in the same way as the planar-diffusion method. When planar-diffusion and epitaxial techniques are combined, the component characteristics are improved because of the uniformity of doping in the epitaxial layer. A cross section of a typical planar-epitaxial transistor is shown in figure 1-13. Note that the component parts do not penetrate the substrate as they did in the planar-diffused transistor.

COLLECTOR CONTACT
EPITAXIAL LAYER

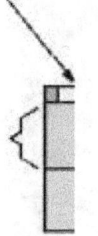

CONTACT
EMITTER CONTACT
/
ii n
I EMITTER~T
BASE
COLLECTOR
SILICON SUBSTRATE
I I OXIDE LAYER || LEAD CONTACTS
Figure 1-13.—Planar-epitaxial transistor.

ISOLATION.—Because of the closeness of components in ICs, ISOLATION from each other becomes a very important factor. Isolation is the prevention of unwanted interaction or leakage between components. This leakage could cause improper operation of a circuit.

Techniques are being developed to improve isolation. The most prominent is the use of silicon oxide, which is an excellent insulator. Some manufacturers are experimenting with single-crystal silicon grown on an insulating substrate. Other processes are also used which are far too complex to go into here. With progress in isolation techniques, the reliability and efficiency of ICs will increase rapidly.

Thin Film

Thin film is the term used to describe a technique for depositing passive circuit elements on an insulating substrate with coating to a thickness of 0.0001 centimeter. Many methods of thin-film deposition exist, but two of the most widely used are VACUUM EVAPORATION and CATHODE SPUTTERING.

VACUUM EVAPORATION.—Vacuum evaporation is a method used to deposit many types of materials in a highly evacuated chamber in which the material is heated by electricity, as shown in figure 1-14. The material is radiated in straight lines in all directions from the source and is shadowed by any objects in its path.

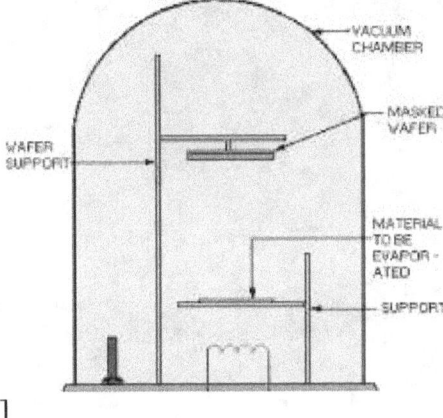

]
-VACUUM PUMP

Figure 1-14.—Vacuum evaporation oven.

The wafers, with appropriate masks (figure 1-15), are placed above and at some distance from the material being evaporated. When the process is completed, the vacuum is released and the masks are removed from the wafers. This process leaves a thin, uniform film of the deposition material on all parts of the wafers exposed by the open portions of the mask. This process is also used to deposit interconnections (leads) between components of an IC.

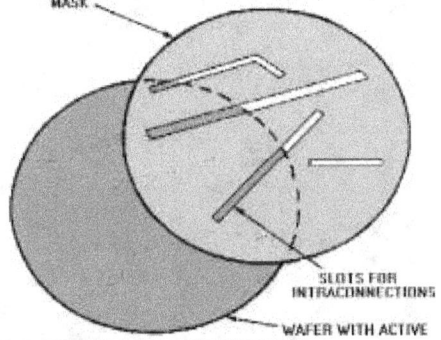

ELEMENTS DIFFUSED

Figure 1-15.—Evaporation mask.

The vacuum evaporation technique is most suitable for deposition of highly reactive materials, such as aluminum, that are difficult to work with in air. The method is clean and allows a better contact between the layer of deposited material and the surface upon which it has been deposited. In addition, because evaporation beams travel in straight lines, very precise patterns may be produced.

CATHODE-SPUTTERING—A typical cathode-sputtering system is illustrated in figure 1-16. This process is also performed in a vacuum. A potential of 2 to 5 kilovolts is applied between the anode and cathode (source material). This produces a GLOW DISCHARGE in the space between the electrodes. The rate at which atoms are SPUTTERED off the source material depends on the number of ions that strike it and the number of atoms ejected for each ion bombardment. The ejected atoms are deposited uniformly over all objects within the chamber. When the sputtering cycle is completed, the vacuum in the chamber is released and the wafers are removed. The masks are then removed from the wafers, leaving a deposit that forms the passive elements of the circuit, as shown in figure 1-17.

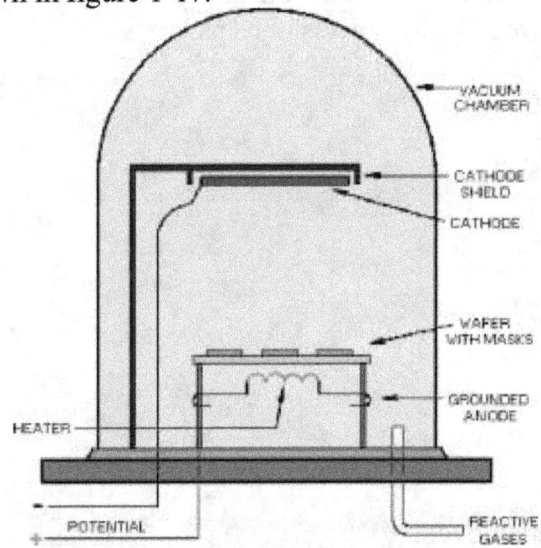

Figure 1-16.—Cathode-sputtering system.

RESISTIVE ELEMENTS

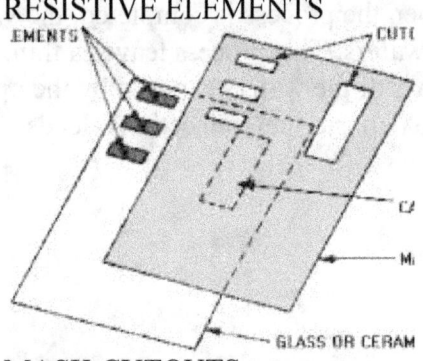

MASK CUTOUTS
CAPACITOR
MASK
GLASS OR CERAMIC SUBSTRATE

Figure 1-17.—Cathode-sputtering mask.

Finely polished glass, glazed ceramic, and oxidized silicon have been used as substrate materials for thin films. A number of materials, including nichrome, a compound of silicon oxide and chromium cermets, tantalum, and titanium, have been used for thin-film resistors. Nichrome is the most widely used.

The process for producing thin-film capacitors involves deposition of a bottom electrode, a dielectric, and finally a top electrode. The most commonly used dielectric materials are silicon monoxide and silicon dioxide.

Thick films are produced by screening patterns of conducting and insulating materials on ceramic substrates. A thick film is a film of material with a thickness that is at least 10 times greater than the mean free path of an electron in that material, or approximately 0.001 centimeter. The technique is used to produce only passive elements, such as resistors and capacitors.

PROCEDURES.—One procedure used in fabricating a thick film is to produce a series of stencils called SCREENS. The screens are placed on the substrate and appropriate conducting or insulating materials are wiped across the screen. Once the conducting or insulating material has been applied, the screens are removed and the formulations are fired at temperatures above 600 degrees Celsius. This process forms alloys that are permanently bonded to the insulating substrate. To a limited extent, the characteristics of the film can be controlled by the firing temperature and length of firing time.

RESISTORS.—Thick-film resistance values can be held to a tolerance of ±10 percent. Closer tolerances are obtained by trimming each resistor after fabrication. Hundreds of different cermet formulations are used to produce a wide range of component parameters. For example, the material used for a 10-ohm-per-square resistor is quite different from that used for a 100-kilohm-per-square resistor.

CAPACITORS AND RESISTOR-CAPACITOR NETWORKS —Capacitors are formed by a sequence of screenings and firings. Capacitors in this case consist of a bottom plate, intraconnections, a dielectric, and a top plate. For resistor-capacitor networks, the next step would be to deposit the resistor material through the screen. The final step is screening and firing of a glass enclosure to seal the unit.

Thick Film

Hybrid Microcircuit

A hybrid microcircuit is one that is fabricated by combining two or more circuit types, such as film and semiconductor circuits, or a combination of one or more circuit types and discrete elements. The primary advantage of hybrid microcircuits is design flexibility; that is, hybrid microcircuits can be designed to provide wide use in specialized applications, such as low-volume and high-frequency circuits.

Several elements and circuits are available for hybrid applications. These include discrete components that are electrically and mechanically compatible with ICs. Such components may be used to perform functions that are supplementary to those of ICs. They can be handled, tested, and assembled with essentially the same technology and tools. A hybrid IC showing an enlarged chip is shown in figure 1-18.

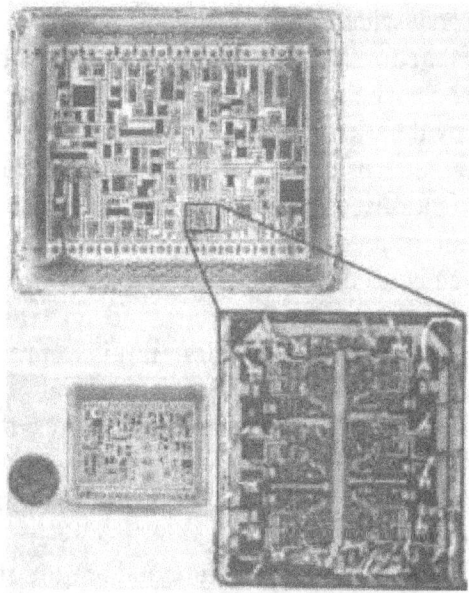

Figure 1-18.—Hybrid IC showing an enlarged chip.

Complete circuits are available in the form of UNCASED CHIPS (UNENCAPSULATED IC DICE). These chips are usually identical to those sold as part of the manufacturer's regular production line. They must be properly packaged and connected by the user if a high-quality final assembly is to be obtained. The circuits are usually sealed in a package to protect them from mechanical and environmental stresses. One-mil (0.001-inch), gold-wire leads are connected to the appropriate pins which are brought out of the package to allow external connections.

Q22. Name the two types of monolithic IC construction discussed.

Q23. How do the two types of monolithic IC construction differ?

Q24. What is isolation?

Q25. What methods are used to deposit thin-film components on a substrate?

Q26. How are thick-film components produced?

Q2 7. What is a hybrid IC?

Q28. What is the primary advantage of hybrid circuits?

PACKAGING TECHNIQUES

Once the IC has been produced, it requires a housing that will protect it from damage. This damage could result from moisture, dirt, heat, radiation, or other sources. The housing protects the device and aids in its handling and connection into the system in which the IC is used. The three most common types of packages are the modified TRANSISTOR-OUTLINE (TO) PACKAGE, the FLAT PACK, and the DUAL INLINE PACKAGE (DIP).

Transitor-Outline Package

Transistor-Outline Package. The transistor-outline (TO) package was developed from early experience with transistors. It was a reliable package that only required increasing the number of leads to make it useful for ICs. Leads normally number between 2 and 12, with 10 being the most common for IC applications. Figure 1-19 is an exploded view of a TO-5 package. Once the IC has been attached to the header, bonding wires are used to attach the IC to the leads. The cover provides the necessary protection for the device. Figure 1-20 is an enlarged photo of an actual TO-5 with the cover removed. You can easily see that the handling of an IC without packaging would be difficult for a technician.

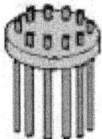

WIRE BONDS BONDING ISLAND MONOLITHIC DIE SOLDER HEADER /

KOVAR EYELET PLATED WITH GOLD CYANIDE PREFORM LEADS

Figure 1-19.—Exploded TO-5.

Figure 1-20.—TO-5 package.

The modified TO-5 package (figure 1-21) can be either plugged into [view (A)] or embedded in [view (B)] a board. The embedding method is preferred. Whether the package is plugged in or embedded, the interconnection area of the package leads must have sufficient clearance on both sides of the board. The plug-in method does not provide sufficient clearance between pads to route additional circuitry. When the packages are embedded, sufficient space exists between the pads [because of the increased diameter of the interconnection pattern, shown at the right in view (B)] for additional conductors.

1-20

(A)PLUG-IN MOUNTING

Figure 1-21A.—TO-5 mounting PLUG-IN MOUNTING

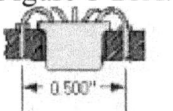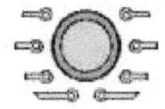

(B)EMBEDDED CAN
(LEADS PLUGGED IN]

Figure 1-21B.—TO-5 mounting EMBEDDED CAN(LEADS PLUGGED IN)

Flat Pack

Many types of IC flat packs are being produced in various sizes and materials. These packages are available in square, rectangular, oval, and circular configurations with 10 to 60 external leads. They may be

made of metal, ceramic, epoxy, glass, or combinations of those materials. Only the ceramic flat pack will be discussed here. It is representative of all flat packs with respect to general package requirements (see figure 1-22).

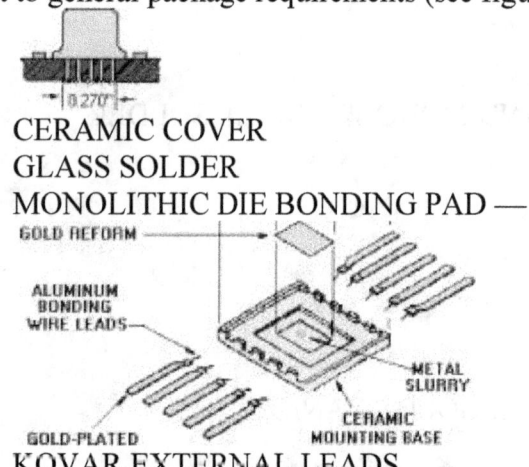

CERAMIC COVER
GLASS SOLDER
MONOLITHIC DIE BONDING PAD —

Figure 1-22.—Enlarged flat pack exploded view.

After the external leads are sealed to the mounting base, the rectangular area on the inside bottom of the base is treated with metal slurry to provide a surface suitable for bonding the monolithic die to the base. The lead and the metalized area in the bottom of the package are plated with gold. The die is then attached by gold-silicon bonding.

The die-bonding step is followed by bonding gold or aluminum wires between the bonding islands on the IC die and on the inner portions of the package leads. Next, a glass-soldered preformed frame is placed on top of the mounting base. One surface of the ceramic cover is coated with Pyroceram glass, and the cover is placed on top of the mounting base. The entire assembly is placed in an oven at 450 degrees Celsius. This causes the glass solder and Pyroceram to fuse and seal the cover to the mounting base. A ceramic flat pack is shown in figure 1-23. It has been opened so that you can see the chip and bonding wires.

Figure 1-23.—Typical flat pack.

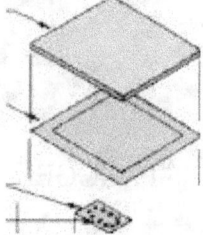

Dual Inline Package

The dual inline package (DIP) was designed primarily to overcome the difficulties associated with handling and inserting packages into mounting boards. DIPs are easily inserted by hand or machine and require no spreaders, spacers, insulators, or lead-forming tools. Standard hand tools and soldering irons can be used to field-service the devices. Plastic DIPs are finding wide use in commercial applications, and a number of military systems are incorporating ceramic DIPS.

The progressive stages in the assembly of a ceramic DIP are illustrated in figure 1-24, views (A) through (E). The integrated-circuit die is sandwiched between the two ceramic elements, as shown in view (A). The element on the left of view (A) is the bottom half of the sandwich and will hold the integrated-circuit die. The ceramic section on the right is the top of the sandwich. The large well in view (B) protects the IC die from mechanical stress during sealing operations. Each of the ceramic elements is coated with glass which has a low melting temperature for subsequent joining and sealing. View (B) shows the Kovar lead frame stamped and bent into its final shape. The excess material is intended to preserve pin alignment. The holes at each end are for the keying jig used in the final sealing operation. The lower half of the ceramic package is inserted into the lead frame shown in view (C). The die is mounted in the well and leads are attached. The top ceramic elements are bonded to the bottom element shown in view (D) and the excess material is removed from the package. View (E) is the final product.

Ceramic DIPs are processed individually while plastic DIPs are processed in quantities of two or more (in chain fashion). After processing, the packages are sawed apart. The plastic package also uses a Kovar lead frame, but the leads are not bent until the package is completed. Because molded plastic is

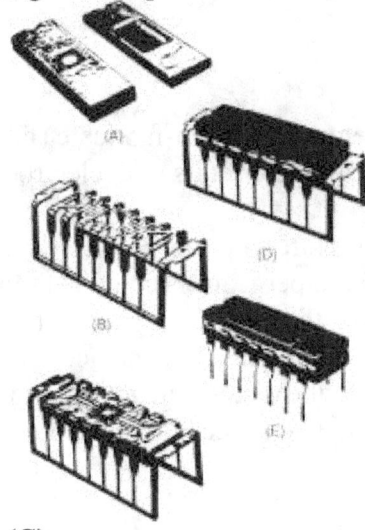

(C)

Figure 1-24.—DIP packaging steps.

used to encapsulate the IC die, no void will exist between the cover and die, as is the case with ceramic packaging.

At present, ceramic DIPs are the most common of the two package types to be found in Navy microelectronic systems. Figure 1-25 shows a DIP which has been opened.

Figure 1-25.—Dual inline package (DIP).

RECENT DEVELOPMENTS IN PACKAGING

Considerable effort has been devoted to eliminating the fine wires used to connect ICs to Kovar leads. The omission of these wires reduces the cost of integrated circuits by eliminating the costs associated with the bonding process. Further, omission of the wires improves reliability by eliminating a common cause of circuit failure.

A promising packaging technique is the face-down (FLIP-CHIP) mounting method by which conductive patterns are evaporated inside the package before the die is attached. These patterns connect the external leads to bonding pads on the inside surface of the die. The pads are then bonded to appropriate pedestals on the package that correspond to those of the bonding pads on the die (figure 1-26).

INVERTED DIE

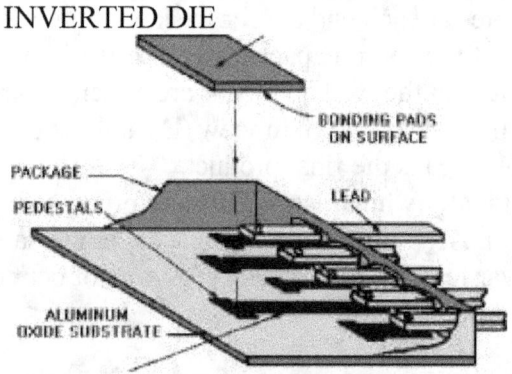

DEPOSITED CONDUCTORS

Figure 1-26.—Flip-chip package.

The BEAM-LEAD technique is a process developed to batch-fabricate (fabricate many at once) semiconductor circuit elements and integrated circuits with electrodes extended beyond the edges of the

wafer, as shown in figure 1-27. This type of structure imposes no electrical difficulty, and parasitic capacitance (under 0.05 picofarad per lead) is equivalent to that of a wire-bonded and brazed-chip assembly. In addition, the electrodes may be tapered to allow for lower inductance, impedance matching, and better heat conductance. The beam-lead technique is easily accomplished and does not have the disadvantages of chip brazing and wire bonding. The feasibility of this technique has been demonstrated in a variety of digital, linear, and thin-film circuits.

BEAM LEAD.

Figure 1-27.—Beam-lead technique.

Another advance in packaging is that of increasing the size of DIPs. General purpose DIPs have from 4 to 16 pins. Because of lsi and vlsi, manufacturers are producing DIPs with up to 64 pins. Although size is increased considerably, all the advantages of the DIP are retained. DIPs are normally designed to a particular specification set by the user.

Q29. What is the purpose of the IC package?

Q30. What are the three most common types of packages?

Q31. What two methods of manufacture are being used to eliminate bonding wires?

EQUIVALENT CIRCUITS

At the beginning of this topic, we discussed many applications of microelectronics. You should understand that these applications cover all areas of modern electronics technology. Microelectronic ICs are produced that can be used in many of these varying circuit applications to satisfy the needs of modern technology. This section will introduce you to some of these applications and will show you some EQUIVALENT CIRCUIT comparisons of discrete components and integrated circuits.

J-K FLIP-FLOP AND IC SIZES

Integrated circuits can be produced that combine all the elements of a complete electronic circuit. This can be done with either a single chip of silicon or a single chip of silicon in combination with film components. The importance of this new production method in the evolution of microelectronics can be demonstrated by comparing a conventional J-K flip-flop circuit incorporating solid-state discrete devices and the same type of circuit employing integrated circuitry. (A J-K flip-flop is a circuit used primarily in computers.)

You should recall from NEETS, Module 13, Introduction to Number Systems, Boolean Algebra, and Logic Circuits, that a basic flip-flop is a device having two stable states and two input terminals (or types of input signals), each of which corresponds to one of the two states. The flip-flop remains in one state until caused to change to the other state by application of an input voltage pulse.

A J-K flip-flop differs from the basic flip-flop because it has a third input terminal. A clock pulse, or trigger, is usually applied to this input to ensure proper timing in the circuit. An input signal must occur at the same time as the clock pulse to change the state of the flip-flop. The conventional J-K flip-flop circuit in figure 1-28 requires approximately 40 discrete components, 200 connections, and 300 processing operations. Each of these 300 operations (seals and connections) represents a possible source of failure. If all the elements of this circuit are integrated into one chip of silicon, the number of connections drops to approximately 14. This is because all circuit elements are intraconnected inside the package and the 300 processing operations are reduced to approximately 30. Figure 1-29 represents a size comparison of a discrete J-K circuit and an integrated circuit of the same type.

SET
CLEAR
INPUTS <

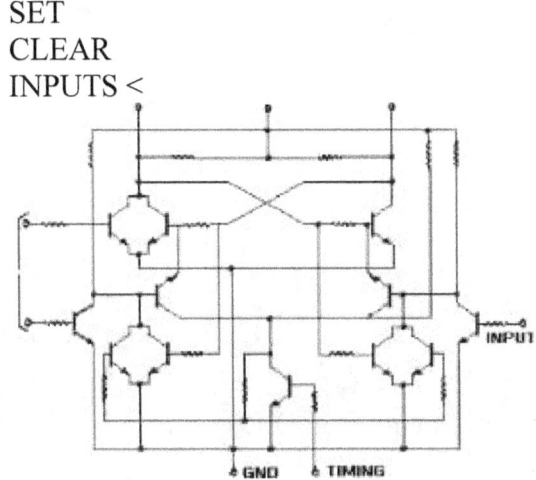

Figure 1-28.—Schematic diagram of a J-K flip-flop.
fl"-
• ^ o • o
o 0 0 o (00 o 0 o

oMS 000 8 (f
innnnnnnnnnnnnnn
(A) DISCHARGE CIRCUIT

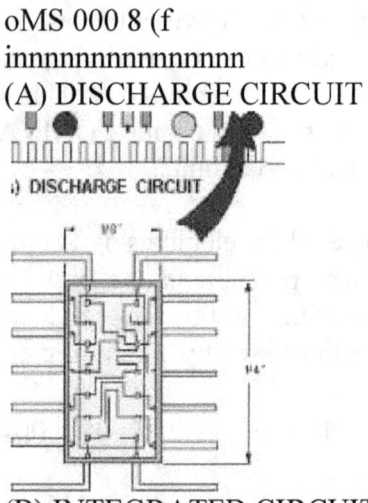

:) DISCHARGE CIRCUIT

(B) INTEGRATED CIRCUIT
Figure 1-29.—J-K flip-flop discrete component and an IC.

IC PACKAGE LEAD IDENTIFICATION (NUMBERING)

When you look at an IC package you should notice that the IC could be connected incorrectly into a circuit. Such improper replacement of a component would likely result in damage to the equipment. For this reason, each IC has a REFERENCE MARK to align the component for placement. The dual inline package (both plastic and ceramic) and the flat pack have a notch, dot, or impression on the package. When the package is viewed from the top, pin 1 will be the first pin in the counterclockwise direction next to the reference mark. Pin 1 may also be marked directly by a hole or notch or by a tab on it (in this case pin 1 is the counting reference). When the package is viewed from the top, all other pins are numbered consecutively in a counterclockwise direction from pin 1, as shown in figure 1-30, views (A) and (B).

m m m m m m m

LsJ LiJ LuU LllI LhJ IhJ LllJ
TOP VIEW
8 7
9 10
T I 5 4 3 *
I 11 12 13 14 >
2 1
15 16
TOP VIEW
(A) DIP
Figure 1-30A.—DIP and flat-pack lead numbering. DIP
TAB
/
n r
2 ∎ 1*13
3 12

4 11

5 ID

6 7 a 8

n

TOP VIEW

(B) Flat-Pack

Figure 1-30B.—DIP and flat-pack lead numbering. Flat-Pack

The TO-5 can has a tab for the reference mark. When numbering the leads, you must view the TO-5 can from the bottom. Pin 1 will be the first pin in a clockwise direction from the tab. All other pins will be numbered consecutively in a clockwise direction from pin 1, as shown in figure 1-31.

BOTTOM VIEW

BOTTOM VIEW

Figure 1-31.—Lead numbering for a TO-5.

IC IDENTIFICATION

As mentioned earlier, integrated circuits are designed and manufactured for hundreds of different uses. Logic circuits, clock circuits, amplifiers, television games, transmitters, receivers, and musical instruments are just a few of these applications.

In schematic drawings, ICs are usually represented by one of the schematic symbols shown in figure 1-32. The IC is identified according to its use by the numbers printed on or near the symbol. That series of numbers and letters is also stamped on the case of the device and can be used along with the data sheet, as shown in the data sheet in figure 1-33, by circuit designers and maintenance personnel. This data sheet is provided by the manufacturer. It provides a schematic diagram and describes the type of device, its electrical characteristics, and typical applications. The data sheet may also show the pin configurations with all pins labeled. If the pin configurations are not shown, there may be a schematic diagram showing pin functions. Some data sheets give both pin configurations and schematic diagrams, as shown in figure 1-34. This figure illustrates a manufacturer's data sheet with all of the pin functions shown.

SN 7400 4(181

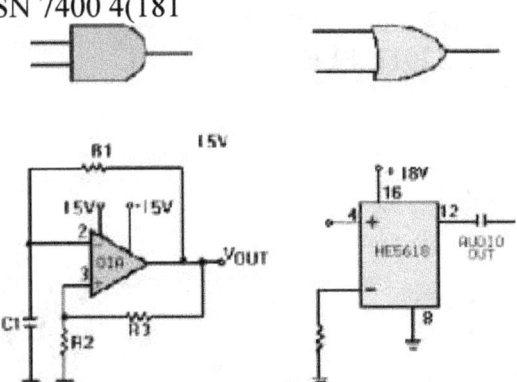

Figure 1-32.—Some schematic symbols for ICs.

FOR AMPLIFIERS. VOLTAGE COMPARATORS. LOW DRIFT SAMPLE AND HOLD

Features

• Low Offsets and Temperature Drift

• Internal 30 pF Capacitor for Frequency Compensation

• Operation from ±5 to ±20 Volt Power Supplies

- Low Current Drain, 1.8 mA at ±20 Volts Typical
- Continuous Short-Circuit Protection
- No Latch Up When Common Mode Range Is Exceeded
- Same Pin Configuration as 709 Amplifier Description

The LH101/LH201 is stable for all feedback configurations, even with capacitive loads, with no external compensation capacitors. Low power dissipation permits high voltage operation across the full temperature range.

PIN CONFIGURATIONS

Maul Can Packaga

Flit Packaga

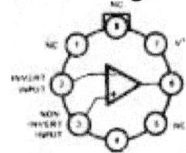

Du*Mn-Lim Package

nc i [

NC |

INVERT INPUT 1 WON

INVERT INPUT

=1

nc cTT

INVERT r yI INPUT S_ WON

iNVf RT fl INPUT

NC cfT

IT^ comp

—f\ H

W3 OUTPUT BALANCE

TOP VIEW NOTE PIN 4 CONNECTED TO CASE

ORDER NUMBERS: LH101H OR LH201H SEE PACKAGE 1

ORDER NUMBERS: LH101F OR LH201F SEE PACKAGE 4

ORDER NUMBERS: LH101D OR LH201D SEE PACKAGE 11

SCHEMATIC DIAGRAM

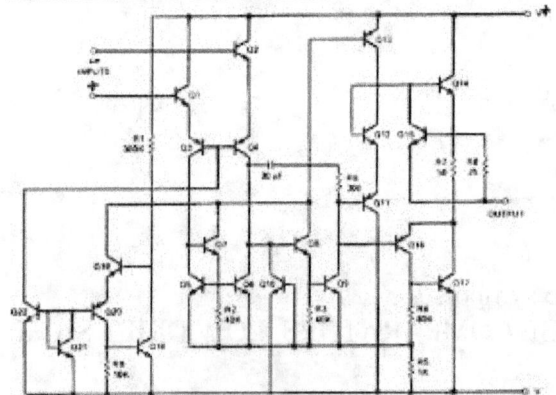

* NC - NOT CONNECTED INTERNALLY

TYPICAL APPLICATIONS

FET Operational Amplifier

Ov'

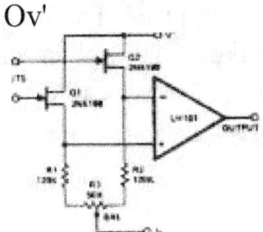

Integrator with Bias Current Compensation

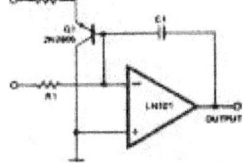

INPUT O vVv*

'SELECT FOR ZERO INTEGRATOR DRIFT.

Courtesy of Siliconix Incorporated

Figure 1-33.—Manufacturer's Data Sheet.

1-30

FEATURES

DESCRIPTION

• MONOLITHIC CONSTRUCTION
• INITIAL ACCURACY
• OUTPUT VOLTAGE ERROR, TOTAL
• LOW NOISE
• WIDE INPUT RANGE
• LOW POWER DISSIPATION
• OUTPUT SHORT CIRCUIT PROTECTION
• ADJUSTABLE OUTPUT

+10V± 0.010V ±1/4 LSB

12V TO 30V 30mW

APPLICATIONS

m AN ECONOMICAL EXTERNAL REFERENCE FOR: HI-5608; OAC OB; AD1408; AD559

• VOLTAGE REGULATOR REFERENCE
• PORTABLE BATTERY OPERATED EQUIPMENT
• NEGATIVE 10V REFERENCE

HA-1608 is a monolithic +10V adjustable voltage reference featuring accuracy and temperature stability specifications detailed exclusively for 8 bit data conversion systems. A stable -(-10V output is provided by a reference zsner and buffer amplifier coupled with laser trimmed feedback and zener bias resistors. Long term stability is ensured through integration of all reference components into a monolithic design. Flexibility of HA-1608 is provided through an external trim control which allows the user to adjust the output voltage for binary or BCD applications without affecting overall performance.

These devices provide a total output voltage error of i 1/4 LSB for 8 bit 0/A or A/D converters. Low standby power (0.3mW) makes HA-1608 a natural selection for portable battery operated equipment, comparator references, and reference stacking circuits. These devices can also be used on -10V references.

HA-1608 is packaged in 8 pin metal cans<TO-99) and 8 pin OIPs. The pinout is arranged for convenient replacement of other less accurate regulators in applications demanding minimal change with temperature and time. HA-1608-2 is specified for -55°C to +125°C operation while the HA-1608-5 operates from 0°C to +75°C.

PINOUT

FUNCTIONAL SCHEMATIC

Section 11 for P»ekaging

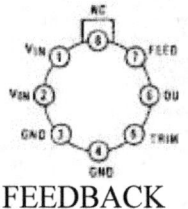

FEEDBACK

OUTPUT

TRIM

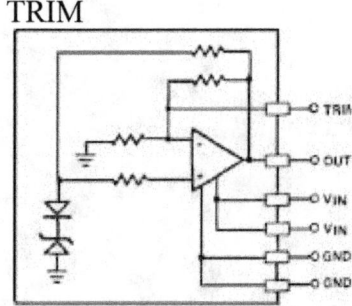

♦ NC - NOT CONNECTED INTERNALLY

Courtesy of Harris Semiconductors

Figure 1-34.—Manufacturer's Data Sheet.

Q32. On DIP and flat-pack ICs viewed from the top, pin I is located on which side of the reference mark?

Q33. DIP and flat-pack pins are numbered consecutively in what direction? Q34. DIP and flat-pack pins are numbered consecutively in what direction? Q35. Viewed from the bottom, TO-5 pins are counted in what direction? Q36. The numbers and letters on ICs and schematics serve what purpose?

MICROELECTRONIC SYSTEM DESIGN CONCEPTS

You should understand the terminology used in microelectronics to become an effective and knowledgeable technician. You should be familiar with packaging concepts from a maintenance standpoint and be able to recognize the different types of assemblies. You should also know the electrical and environmental factors that can affect microelectronic circuits. In the next section of this topic we will define and discuss each of these areas.

TERMINOLOGY

As in any special electronics field, microelectronics terms and definitions are used to clarify communications. This is done so that everyone involved in microelectronics work has the same knowledge of the field. You can imagine how much trouble you would have remembering 10 or more different names and definitions for a resistor. If standardization didn't exist for the new terminology, you would have far more trouble understanding microelectronics. To standardize terminology in microelectronics, the

Navy has adopted several definitions with which you should become familiar. These definitions will be presented in this section.

Microelectronics

Microelectronics is that area of electronics technology associated with electronics systems built from extremely small electronic parts or elements. Most of today's computers, weapons systems, navigation systems, communications systems, and radar systems make extensive use of microelectronics technology.

Microcircuit

A microcircuit is not what the old-time technician would recognize as an electronic circuit. The old-timer would no longer see the familiar discrete parts (individual resistors, capacitors, inductors, transistors, and so forth). Microelectronic circuits, as discussed earlier, are complete circuits mounted on a substrate (integrated circuit). The process of fabricating microelectronic circuits is essentially one of building discrete component characteristics either into or onto a single substrate. This is far different from soldering resistors, capacitors, transistors, inductors, and other discrete components into place with wires and lugs. The component characteristics built into microcircuits are referred to as ELEMENTS rather than discrete components. Microcircuits have a high number of these elements per substrate compared to a circuit with discrete components of the same relative size. As a matter of fact, microelectronic circuits often contain thousands of times the number of discrete components. The term HIGH EQUIVALENT CIRCUIT DENSITY is a description of this element-to-discrete part relationship. For example, suppose you have a circuit with 1,000 discrete components mounted on a chassis which is 8 x 10 x 2 inches. The equivalent circuit in microelectronics might be built into or onto a single substrate which is only 3/8 x 1 x 1/4 inch. The 1,000 elements of the microcircuit would be very close to each other (high density) by

comparison to the distance between discrete components mounted on the large chassis. The elements within the substrate are interconnected on the single substrate itself to perform an electronic function. A microcircuit does not have any discrete components mounted on it as do printed circuit boards, circuit card assemblies, and modules composed exclusively of discrete component parts.

Microcircuit Module

Microcircuits may be used in combination with discrete components. An assembly of microcircuits or a combination of microcircuits and discrete conventional electronic components that performs one or more distinct functions is a microcircuit module. The module is constructed as an independently packaged, replaceable unit. Examples of microcircuit modules are printed circuit boards and circuit card assemblies. Figure 1-35 is a photograph of a typical microcircuit module.

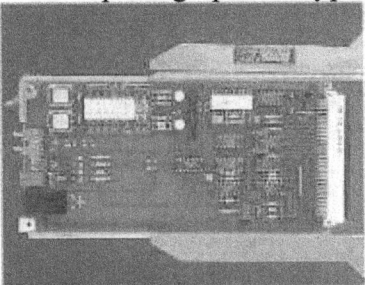

Figure 1-35.—Microcircuit module.

Miniature Electronics

Miniature electronics includes miniature electronic components and packages. Some examples are printed circuit boards, printed wiring boards, circuit card assemblies, and modules composed exclusively of discrete electronic parts and components (excluding microelectronic packages) mounted on boards, assemblies, or modules. MOTHER BOARDS, large printed circuit boards with plug-in modules, are considered miniature electronics. Cordwood modules also fall into this category. Miniature motors, synchros, switches, relays, timers, and so forth, are also classified as miniature electronics.

Recall that microelectronic components contain integrated circuits. Miniature electronics contain discrete elements or parts. You will notice that printed circuit boards and circuit card assemblies are mentioned in more than one definition. To identify the class (microminiature or miniature) of the unit, you must first determine the types of components used.

Q37. Standardized terms improve what action between individuals?

Q38. Microcircuit refers to any component containing what types of elements?

Q39. Components made up exclusively of discrete elements are classified as what type of electronics?

SYSTEM PACKAGING

When a new electronics system is developed, several areas of planning require special attention. An area of great concern is that of ensuring that the system performs properly. The designer must take into

account all environmental and electrical factors that may affect the system. This includes temperature, humidity, vibration, and electrical interference. The design factor that has the greatest impact on you, as the technician, is the MAINTAINABILITY of the system. The designer must take into account how well you will be able to locate problems, identify the faulty components, and make the necessary repairs. If a system cannot be maintained easily, then it is not an efficient system. PACKAGING, the method of enclosing and mounting components, is of primary importance in system maintainability.

Levels of Packaging

For the benefit of the technician, system packaging is usually broken down to five levels (0 to IV). These levels are shown in figure 1-36.

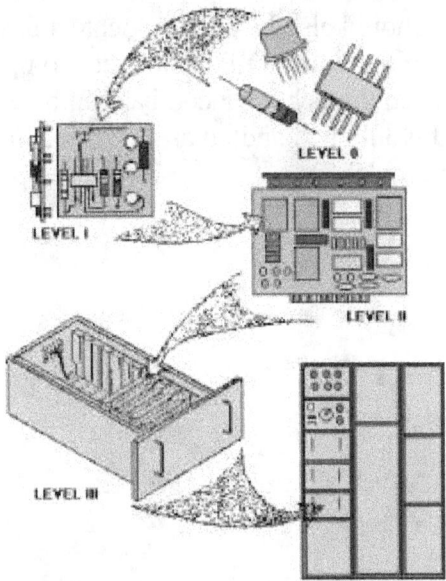

LEVEL IV

Figure 1-36.—Packaging levels.

LEVEL 0.—Level 0 packaging identifies nonrepayable parts, such as integrated circuits, transistors, resistors, and so forth. This is the lowest level at which you can perform maintenance. You are limited to simply replacing the faulty element or part. Depending on the type of part, repair might be as simple as plugging in a new relay. If the faulty part is an IC, special training and equipment will be required to accomplish the repair. This will be discussed in topic 2.

LEVEL I.—This level is normally associated with small modules or submodules that are attached to circuit cards or mother boards. The analog-to-digital (A/D) converter module is a device that converts a signal that is a function of a continuous variable (like a sine wave) into a representative number sequence in digital form. The A/D converter in figure 1-37 is a typical Level I component. At this level of

maintenance you can replace the faulty module with a good one. The faulty module can then be repaired at a later time or discarded. This concept significantly reduces the time equipment is inoperable.

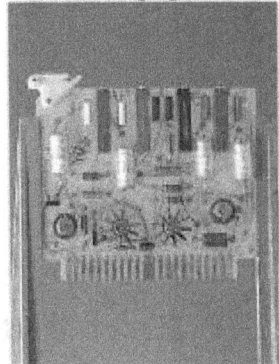

Figure 1-37.—Printed circuit board (pcb).

LEVEL II.—Level II packaging is composed of large printed circuit boards and/or cards (mother boards). Typical units of this level are shown in figures 1-37 and 1-38. In figure 1-38 the card measures 15 x 5.25 inches. The large dual inline packages (DIPs) are 2.25 inches x 0.75 inch. Other DIPs on the pcb are much smaller.

Interconnections are shown between DIPs. You should also be able to locate a few discrete components. Repair consists of removing the faulty DIP or discrete component from the pcb and replacing it with a new part. Then the pcb is placed back into service. The removed part may be a level 0 or I part and would be handled as described in those sections. In some cases, the entire pcb should be replaced.

Figure 1-38.—Printed circuit board (pcb).

LEVEL III.—Drawers or pull-out chassis are level III units, as shown in figure 1-36. These are designed for accessibility and ease of maintenance. Normally, circuit cards associated with a particular subsystem will be grouped together in a drawer. This not only makes for an orderly arrangement of subsystems but also eliminates many long wiring harnesses. Defective cards are removed from such drawers and defective components are repaired as described in level II.

LEVEL IV.—Level IV is the highest level of packaging. It includes the cabinets, racks, and wiring harnesses necessary to interconnect all of the other levels. Other pieces of equipment of the same system classified as level IV, such as radar antennas, are broken down into levels 0 to III in the same manner.

During component troubleshooting procedures, you progress from level IV to III to II and on to level 0 where you identify the faulty component. As you become more familiar with a system, you should be able to go right to the drawer or module causing the problem.

Q40. Resistors, capacitors, transistors, and the like, are what level of packaging?

Q41. Modules or submodules attached to a mother board are what packaging level?

Q42. What is the packaging level of a pcb?

INTERCONNECTIONS IN PRINTED CIRCUIT BOARDS

As electronic systems become more complex, interconnections between components also becomes more complex. As more components are added to a given space, the requirements for interconnections become extremely complicated. The selection of conductor materials, insulator materials, and component physical size can greatly affect the performance of the circuit. Poor choices of these materials can contribute to poor signals, circuit noise, and unwanted electrical interaction between components. The three most common methods of interconnection are the conventional pcb, the multilayer pcb, and the modular assembly. Each of these will be discussed in the following sections.

Conventional Printed Circuit Board

Printed circuit boards were discussed earlier in topic 1. You should recall that a conventional pcb consists of glass-epoxy insulating base on which the interconnecting pattern has been etched. The board may be single- or double-sided, depending on the number of components mounted on it. Figures 1-37 and 1-38 are examples of conventional printed circuit boards.

Multilayer Printed Circuit Board.

The multilayer printed circuit board is emerging as the solution is interconnection problems associated with high-density packaging. Multilayer boards are used to:

• reduce weight
• conserve space in interconnecting circuit modules
• eliminate costly and complicated wiring harnesses
• provide shielding for a large number of conductors
• provide uniformity in conductor impedance for high-speed switching systems
• allow greater wiring density on boards

Figure 1-39 illustrates how individual boards are mated to form the multilayer unit. Although all multilayer boards are similarly constructed, various methods can be used to interconnect the circuitry from layer to layer. Three proven processes are the clearance-hole, plated-through hole, and layer buildup methods.

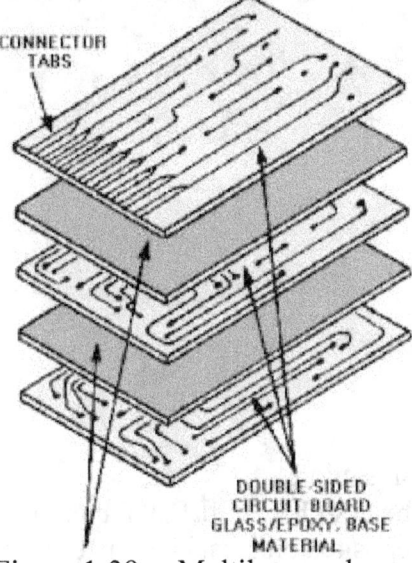

Figure 1-39.—Multilayer pcb.

CLEARANCE-HOLE METHOD.—In the CLEARANCE-HOLE method, a hole is drilled in the copper island (terminating end) of the appropriate conductor on the top layer. This provides access to a conductor on the second layer as shown by hole A in figure 1-40. The clearance hole is filled with solder to complete the connection. Usually, the hole is drilled through the entire assembly at the connection site. This small hole is necessary for the SOLDER-FLOW PROCESS used with this interconnection method.

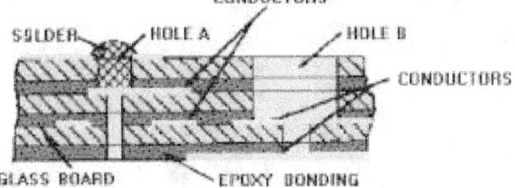

Figure 1-40.—Clearance-hole interconnection.

Conductors located several layers below the top are connected by using a

STEPPED-DOWN HOLE PROCESS. Before assembly of a three-level board, a clearance hole is drilled down to the first layer to be interconnected. The first layer to be interconnected is predrilled with a hole smaller than those drilled in layers 1 and 2; succeeding layers to be connected have progressively smaller clearance holes. After assembly, the exposed portion of the conductors are interconnected by filling the stepped-down holes with solder, as shown by hole B in figure 1-40. The larger the number of interconnections required at one point, the larger must be the diameter of the clearance holes on the top layer. Large clearance holes on the top layer allow less space for components and reduce packaging density.

PLATED-THROUGH-HOLE METHOD.—The PLATED-THROUGH-HOLE method of interconnecting conductors is illustrated in figure 1-41. The first step is to temporarily assemble all the layers into their final form. Holes corresponding to required connections are drilled through the entire assembly and then the unit is disassembled. The internal walls of those holes to be interconnected are plated with metal which is 0.001 inch thick. This, in effect, connects the conductor on the board surface through the hole itself. This process is identical to that used for standard printed circuit boards. The boards are then reassembled and permanently bonded together with heat and pressure. All the holes are plated through with metal.

PRINTED CONDUCTORS
EPOXY GLASS (FDR BONDING)
PRINTED WIRING BOARD

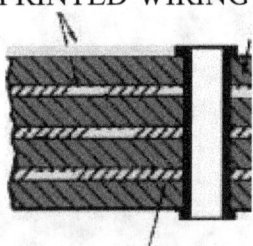

PLATED HOLE
3S
Figure 1-41.—Plated through-hole interconnection.

LAYER BUILD-UP METHOD.—With the LAYER BUILD-UP method, conductors and insulation layers are alternately deposited on a backing material, as shown in figure 1-42. This method produces copper interconnections between layers and minimizes the thermal expansion effects of dissimilar materials. However, reworking the internal connections in built-up layers is usually difficult, if not impossible.

BACKING MATERIAL

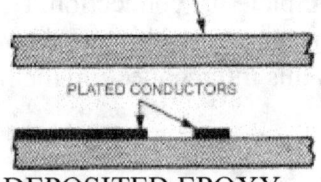

PLATED CONDUCTORS

DEPOSITED EPOXY.
WINDOW
PLATED CONDUCTORS

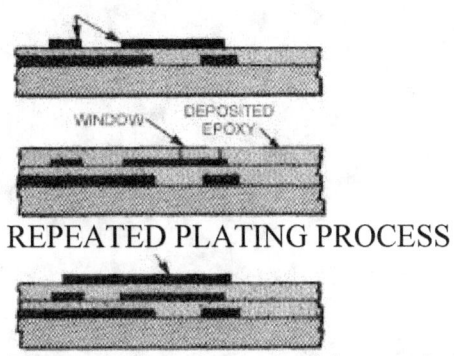

REPEATED PLATING PROCESS

Figure 1-42.—Layer build-up technique.

Advantages and Disadvantages of Printed Circuit Boards

Some of the advantages and disadvantages of printed circuit boards were discussed earlier in this topic. They are strong, lightweight, and eliminate point-to-point wiring. Multilayer printed circuit boards allow more components per card. Entire circuits or even subsystems may be placed on the same card. However, these cards do have some drawbacks. For example, all components are wired into place, repair of cards requires special training and/or special equipment, and some cards cannot be economically repaired because of their complexity (these are referred to as THRO WAWAYS).

MODULAR ASSEMBLIES

The MODULAR-ASSEMBLY (nonrepairable item) approach was devised to achieve ultra-high density packaging. The evolution of this concept, from discrete components to microelectronics, has progressed through various stages. These stages began with cordwood assemblies and functional blocks and led to complete subsystems in a single package. Examples of these configurations are shown in figure 1-43, view (A), view (B), and view (C).

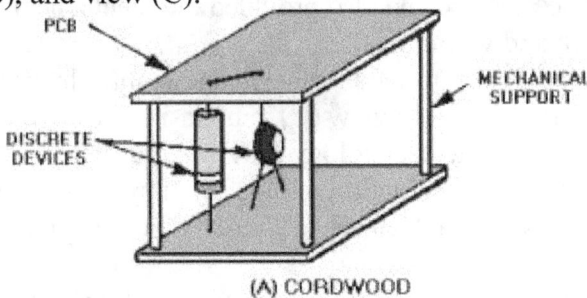

(A) CORDWOOD

Figure 1-43A.—Evolution of modular assemblies. CORDWOOD.

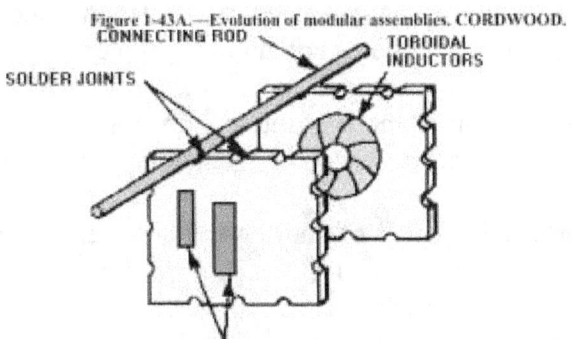

THIN-OR THICK-FILM ELEMNETS (B) MICROMODULE

Figure 1-43B.—Evolution of modular assemblies. MICROMODULE.

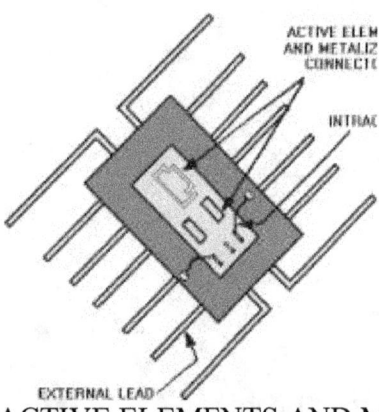

ACTIVE ELEMENTS AND METALIZATION CONNECTORS
INTRACONNECTION
(C) INTEGRATED CIRCUIT
Figure 1-43C—Evolution of modular assemblies. INTEGRATED-CIRCTJIT.
Cordwood Modules.

The cordwood assembly, shown in view (A) of figure 1-43, was designed and fabricated in various forms and sizes, depending on user requirements. This design was used to reduce the physical size and increase the component density and complexity of circuits through the use of discrete devices. However, the use of the technique was somewhat limited by the size of available discrete components used.

Micromodules

The next generation assembly was the micromodule. Designers tried to achieve maximum density in this design by using discrete components, thick- and thin-film technologies, and the insulator substrate principle. The method used in this construction technique allowed for the efficient use of space and also provided the mechanical strength necessary to withstand shock and vibration.

Semiconductor technology was then improved further with the introduction of the integrated circuit. The flat-pack IC form, shown in view (C), emphasizes the density and complexity that exists with this technique. This technology provides the means of reducing the size of circuits. It also allows the reduction of the size of systems through the advent of the lsi circuits that are now available and vlsi circuits that are being developed by various IC manufacturers.

Continuation of this trend toward microminiaturization will result in system forms that will require maintenance personnel to be specially trained in maintenance techniques to perform testing, fault isolation, and repair of systems containing complex miniature and microminiature circuits.

Q43. What are the three most common methods of interconnections?

Q44. Name the three methods of interconnecting components in multilayer printed circuit boards.

Q45. What is one of the major disadvantages of multilayer printed circuit boards?

Q46. What was the earliest form of micromodule?

ENVIRONMENTAL CONSIDERATIONS

The environmental requirements of each system design are defined in the PROCUREMENT SPECIFICATION. Typical environmental requirements for an IC, for example, are shown in table 1-1. After these system requirements have been established, components, applications, and packaging forms are considered. This then leads to the most effective system form.

Table 1-1.—Environmental Requirements

In the example in table 1-1, the environmental requirements are set forth as MILITARY STANDARDS for performance. The actual standard for a particular factor is in parentheses. To meet each of these standards, the equipment or component must perform adequately within the test guidelines. For example, to pass the shock test, the component must withstand a shock of 250 to 600 Gs (force of gravity). During vibration testing, the component must withstand vibrations of 5 to 15 cycles per second for 0.06 day, or about 1 1/2 hours; 16 to 25 cycles for 1 hour; and 26 to 33 cycles for 1/2 hour. Rf interference between 30 hertz and 40 gigahertz must not affect the performance of the component. Temperature and humidity factors are self-explanatory.

When selecting the most useful packaging technique, the system designer must consider not only the environmental and electrical performance requirements of the system, but the maintainability aspects as well. The system design will, therefore, reflect performance requirements of maintenance and repair personnel.

ELECTRICAL CONSIDERATIONS

The electrical characteristics of a component can sometimes be adversely affected when it is placed in a given system. This effect can show up as signal distortion, an improper timing sequence, a frequency shift, or numerous other types of unwanted interactions. Techniques designed to minimize the effects of system packaging on component performance are incorporated into system design by planners. These techniques should not be altered during your maintenance. Several of the techniques used by planners are discussed in the following sections.

Ground Planes and Shielding.

At packaging levels I and II, COPPER PLANES with voids, where feed-through is required, can be placed anywhere within the multilayer board. These planes tend to minimize interference between circuits and from external sources.

At other system levels, CROSS TALK (one signal interfering with another), rf generation within the system, and external interference are suppressed through the use of various techniques. These techniques are shown in figure 1-44. As shown in the figure, rf shielding is used on the mating surfaces of the package, cabling is shielded, and heat sinks are provided.

Interconnection and Intraconnections

To meet the high-frequency characteristics and propagation timing required by present and future systems, the device package must not have excessive distributed capacitance and/or inductance. This type of packaging is accomplished in the design of systems using ICs and other microelectronic devices by using shorter leads internal to the package and by careful spacing of complex circuits on printed circuit boards. To take advantage of the inherent speed of the integrated circuit, you must keep the signal propagation time between circuits to a minimum. The signal is delayed approximately 1 nanosecond per foot, so reducing the distance between circuits as much as possible is necessary. This requires the use of structures, such as high-density digital systems with an emphasis on large-scale integration, for systems in the future. Also, maintenance personnel should be especially concerned with the spacing of circuits, lead dress, and surface cleanliness. These factors affect the performance of high-speed digital and analog circuits.

Q47. In what publication are environmental requirements for equipment defined? Q48. In what publication would you find guidelines for performance of military electronic parts? Q49. Who is responsible for meeting environmental and electrical requirements of a system? Q50. What methods are used to prevent unwanted component interaction?

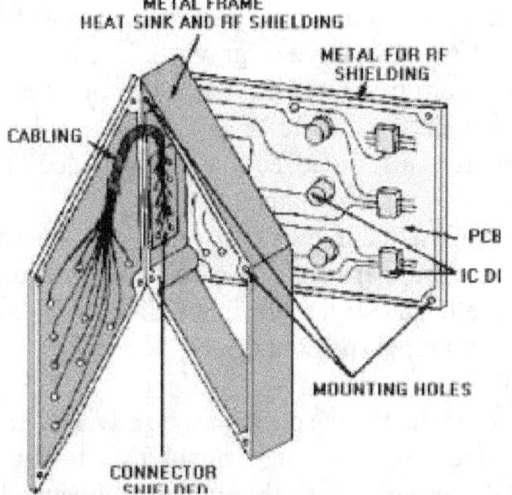

CONNECTOR SHIELDED INPUT AND OUTPUT
IC DEVICES

Figure 1-44.—Ground planes and shielding.

SUMMARY

This topic has presented information on the development and manufacture of microelectronic devices. The information that follows summarizes the important points of this topic.

VACUUM-TUBE CIRCUITS in most modern military equipment are unacceptable because of size, weight, and power use.

Discovery of the transistor in 1948 marked the beginning of MICROELECTRONICS.

The PRINTED CIRCUIT BOARD (pcb) reduces weight and eliminates point-to-point wiring.

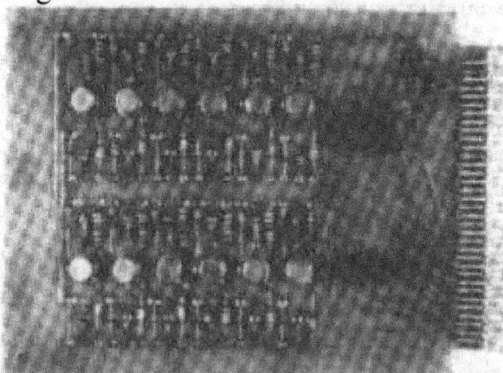

The INTEGRATED CIRCUITS (IC) consist of elements inseparably associated and formed on or within a single SUBSTRATE.

ICs are classified as three types: MONOLITHIC, FILM, and HYBRID.

The MONOLITHIC IC, called a chip or die, contains both active and passive elements.

FILM COMPONENTS are passive elements, either resistors or capacitors.

HYBRID ICs are combinations of monolithic and film or of film and discrete components, or any combination thereof. They allow flexibility in circuits.

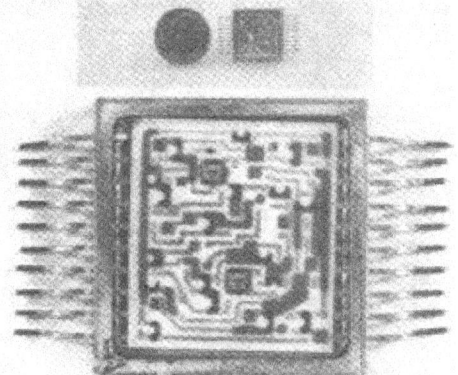

Rapid development has resulted in increased reliability and availability, reduced cost, and higher element density.

LARGE-SCALE (lsi) and VERY LARGE-SCALE INTEGRATION (vlsi) allow thousands of elements in a single chip.

MONOLITHIC ICs are produced by the diffusion or epitaxial methods.

DIFFUSED elements penetrate the substrate, EPITAXIAL do not.

COLLECTOR-N

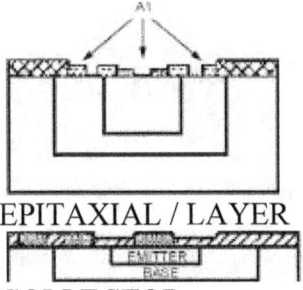

EPITAXIAL / LAYER

COLLECTOR

SILICON SUBSTRATE

ISOLATION is a production method to prevent unwanted interaction between elements within a

chip.

THIN-FILM ELEMENTS are produced through EVAPORATION or CATHODE SPUTTERING techniques.

THICK-FILM ELEMENTS are screened onto the substrate.

The most common types of packages for ICs are TO, FLAT PACK, and DUAL INLINE.

FLIP CHIPS and BEAM-LEAD CHIPS are techniques being developed to eliminate bonding wires and to improve packaging.

INVERTED DIE

BONDING PADS □ N SURFACE

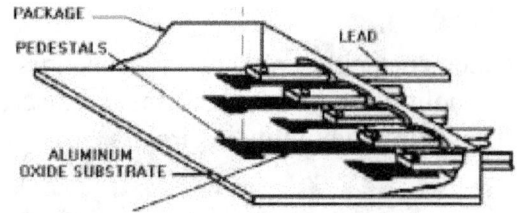

DEPOSITED CONDUCTORS

BEAM LEAD

Large DIPs are being used to package lsi and vlsi. They can be produced with up to 64 pins and are designed to fulfill a specific need.

Viewed from the tops, DIPS and FLAT-PACK LEADS are numbered counterclockwise from the reference mark.

Viewed from the bottom, TO-5 LEADS are numbered clockwise from the tab.

mm m m m m m

L*J L±J h°J LlU [izj LsJ IhJ

TOP VIEW

TOP VIEW (A) DIP

TOP VIEW .TAB

TOP VIEW

(B) FLAT-PACK

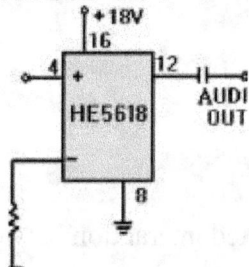

BOTTOM VIEW BOTTOM VIEW

Numbers and letters on schematics and ICs identify the TYPE OF IC.

AUDIO OUT

Knowledge of TERMINOLOGY used in microelectronics and of packaging concepts will aid you in becoming an effective technician.

STANDARD TERMINOLOGY has been adopted by the Navy to ease communication.

MICROELECTRONICS is that area of technology associated with electronic systems designed with extremely small parts or elements.

A MICROCIRCUIT is a small circuit which is considered as a single part composed of elements on or within a single substrate.

A MICROCIRCUIT MODULE is an assembly of microcircuits or a combination of microcircuits and discrete components packaged as a replaceable unit.

MINIATURE ELECTRONICS are card assemblies and modules composed exclusively of discrete electronic components.

SYSTEM PACKAGING refers to the design of a system, taking into account environmental and electronic characteristics, access, and maintainability.

PACKAGING LEVELS 0 to IV are used to identify assemblies within a system. Packaging levels are as follows:

LEVEL O-Nonrepairable parts (resistors, diodes, and so forth.) LEVEL I - Submodules attached to circuit cards. LEVEL II -Circuit cards and MOTHER BOARDS.

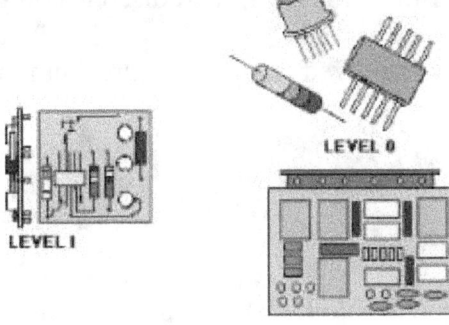

nnr
LEVEL II
LEVEL III - Drawers.

LEVEL IV - Cabinets.
LEVEL IV

The most common METHODS OF INTERCONNECTION are the conventional pcb, the multilayer pcb, and modular assemblies.

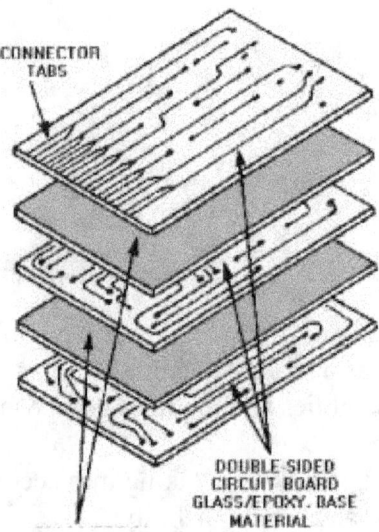

Three methods of interconnecting circuitry in multilayer printed circuit boards are the CLEARANCE-HOLE, the PLATED-THROUGH-HOLE, and LAYER BUILD-UP.

MODULAR ASSEMBLIES were devised to achieve high circuit density.

Modular assemblies have progressed from CORDWOOD MODULES through MICROMODULES. Micromodules consist of film components and discrete components to integrated and hybrid circuitry.

ENVIRONMENTAL FACTORS to be considered are temperature, humidity, shock, vibration, and rf interference.

BACKING MATERIAL

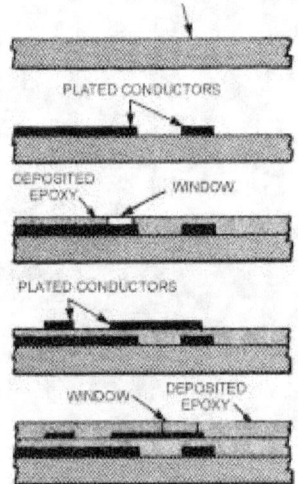

REPEATED PLATING PROCESS

ELECTRICAL FACTORS are overcome by using shielding and ground planes and by careful placement of components.

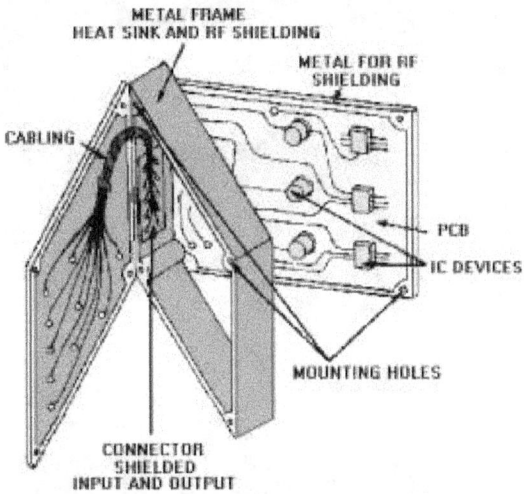

METAL FRAME
HEAT SINK AND RF SHIELDING

METAL FOR RF
SHIELDING

CABLING

PCB

IC DEVICES

MOUNTING HOLES

CONNECTOR
SHIELDED
INPUT AND OUTPUT

ANSWERS TO QUESTIONS Ql. THROUGH Q50.

Al. Size, weight, and power consumption.

A2. The transistor and solid-state diode.

A3. Technology of electronic systems made of extremely small electronic parts or elements.

A4. The Edison Effect.

A5. Transformers, capacitors, and resistors.

A6. "Rat's nest" appearance and unwanted interaction, such as capacitive and inductive effects.

A 7. Rapid repair of systems and improved efficiency.

A8. Differences in performance of tubes of the same type.

A9. Eliminate heavy chassis and point-to-point wiring.

A10. Components soldered in place.

All. Cordwoodmodule.

All. Elements inseparably associated and formed in or on a single substrate.

A13. Monolithic, film, and hybrid.

A14. Monolithic ICs contain active and passive elements. Film ICs contain only passive elements.

A15. Combination of monolithic ICs and film components.

A16. 1,000 to 2,000.

A17. Circuit design, component placement, suitable substrate, and depositing proper materials on substrate.

A18. Complex.

A19. Control patterns of materials on substrates.

A20. Glass or ceramic.

A21. Crystal is sliced into wafers. Then ground and polished to remove any surface defect.

A22. Diffusion; epitaxial growth.

A23. Diffusion penetrates substrate; epitaxial does not.

A24. Electrical separation of elements.

A25. Evaporation and cathode sputtering.

A26. Screening.

A27. Combination of monolithic and film elements.

A28. Circuit flexibility.

A29. Protect the IC from damage; make handling easier.

A30. TO, fiat pack, DIP.

A31. Flip-chip, beam lead.

A32. Left.

A33. Counterclockwise.

A34. Reference mark.

A35. Clockwise.

A36. Identify the type of IC.

A3 7. Communication.

A38. Integrated circuits.

A39. Miniature.

A40. Level 0.

A41. Level I.

A42. Level II.

A43. Conventional printed circuit boards, multilayer printed circuit boards and modular assemblies.

A44. Clearance hole, plated-through hole, and layer build-up.

A45. Difficulty of repair of internal connections.

A46. Cordwood modules.

A47. Procurement specifications.

A48. Military Standards.

A49. Equipment designers (planners).

A50. Ground planes, shielding, component placement.

CHAPTER 2

MINIATURE/MICROMINIATURE (2M) REPAIR PROGRAM AND HIGH-RELIABILITY SOLDERING

LEARNING OBJECTIVES

Upon completion of this topic, the student will be able to:

1. State the purpose and need for training and certification of 2M repair technicians.

2. Explain the maintenance levels at which maintenance is performed.

3. Identify the specialized and general test equipment used in fault isolation.

4. Recognize the specialized types of tools used and the importance of repair facilities.

5. Explain the principles of high-reliability soldering.

INTRODUCTION

As mentioned in topic 1, advances in the field of microelectronics are impressive. With every step forward in production development, a corresponding step forward must be made in maintenance and repair techniques.

This topic will teach you how the Navy is coping with the new technology and how personnel are trained to carry out the maintenance and repair of complex equipment. The program discussed in this topic is up to date at this time, but as industry advances, so must the capabilities of the technician.

MINIATURE AND MICROMINIATURE (2M) ELECTRONIC REPAIR PROGRAM

Training requirements for miniature and microminiature repair personnel were developed under guidelines established by the Chief of Naval Operations. The Naval Sea Systems Command (NAVSEA) developed a program which provides for the proper training in miniature and microminiature repair. This program, NAVSEA Miniature/Microminiature (2M) Electronic Repair, authorizes and provides proper tools and equipment and establishes a personnel certification program to maintain quality repair.

The Naval Air Systems Command has developed a similar program specifically for the aviation community. The two programs are patterned after the National Aeronautics and Space Administration (NASA) high-reliability soldering studies and have few differences other than the administrative chain of command. For purposes of this topic, we will use the NAVSEA manual for reference.

The 2M program covers all phases of miniature and microminiature repair. It establishes the training curriculum for repair personnel, outlines standards of workmanship, and provides guidelines for specific repairs to equipment, including the types of tools to use. This part of the program ensures high-reliability repairs by qualified technicians.

Upon satisfactory completion of a 2M training course, a technician will be CERTIFIED to perform repairs. The CERTIFICATION is issued at the level at which the technician qualifies and specifies what type of repairs the technician is permitted to perform. The two levels of qualification for technicians are MINIATURE COMPONENT REPAIR and MICROMINIATURE COMPONENT REPAIR. Miniature component repair is limited to discrete components and single- and double-sided printed circuit boards, including removal and installation of most integrated circuit devices. Microminiature component repair consists of repairs to highly complex, densely packaged, multilayer printed circuit boards. Sophisticated repair equipment is used that may include a binocular microscope.

To ensure that a technician is maintaining the required qualification level, periodic evaluations are conducted. By inspecting and evaluating the technician's work, certification teams ensure that the minimum standards for the technician's level of qualification are met. If the standards are met, the technician is recertified; if not, the certification is withheld pending retraining and requalification. This portion of the program ensures the high-quality, high-reliability repairs needed to meet operational requirements.

Q1. Training requirements for (2M) repair personnel were developed under guidelines established by what organization?

Q2. What agencies provide training, tools, equipment, and certification of the 2M system?

Q3. To perform microminiature component repair, a 2M technician must be currently certified in what area?

Q4. Multilayer printed circuit board repair is the responsibility of what 2M repair technician?

LEVELS OF MAINTENANCE

Effective maintenance and repair of microelectronic devices require one of three levels of maintenance. Level-of-repair designations called SOURCE, MAINTENANCE, and RECOVERABILITY CODES (SM&R) have been developed and are assigned by the

Chief of Naval Material. These codes are D for DEPOT LEVEL, I for INTERMEDIATE LEVEL, and O for ORGANIZATIONAL LEVEL.

DEPOT-LEVEL MAINTENANCE.

SM&R Code D maintenance is the responsibility of maintenance activities designated by the systems command (NAVSEA, NAVAIR, NAVELEX). This code augments stocks of serviceable material. It also supports codes I and O activities by providing more extensive shop facilities and equipment and more highly skilled technicians. Code D maintenance includes repair, modification, alteration, modernization, and overhaul as well as reclamation or reconstruction of parts, assemblies, subassemblies, and components. Finally, it includes emergency manufacture of nonavailable parts. Code D maintenance also provides technical assistance to user activities and to code I maintenance organizations. Code D maintenance is performed in shops, located in shipyards and shore-based facilities, including contractor maintenance organizations.

INTERMEDIATE-LEVEL MAINTENANCE

SM&R code I maintenance, performed at mobile shops, tenders or shore-based repair facilities (SIMAS) provides direct support to user organizations. Code I maintenance includes calibration, repair, or replacement of damaged or unserviceable parts, components, or assemblies, and emergency manufacture of nonavailable parts. It also provides technical assistance to ships and stations.

ORGANIZATIONAL-LEVEL MAINTENANCE

SM&R code O maintenance is the responsibility of the activity who owns the equipment. Code O maintenance consists of inspecting, servicing, lubricating, adjusting, and replacing parts, minor assemblies, and subassemblies.

An INTEGRATED LOGISTICS SUPPORT PLAN (ILSP) determines the maintenance level for electronic assemblies, modules, and boards for each equipment assigned to an activity. The ILSP codes the items according to the normal maintenance capabilities of that activity. This results in two additional repair-level categories - NORMAL and EMERGENCY.

Normal Repairs

Generally, 2M repairs are performed at the level set forth in the maintenance plan and specified by the appropriate SM&R coding for each board or module. Therefore, normal repairs include all repairs except organizational-level repair of D- and I-coded items and intermediate-level repair of D-coded items.

Emergent/Emergency Repairs

In the NAVSEA 2M Electronic Repair Program, emergent/emergency repairs are those arising unexpectedly. They may require prompt repair action to restore a system or piece of equipment to operating condition where normal repairs are not authorized. These Code O repairs on boards or modules are normally SM&R-coded for Code D repairs. Emergent/emergency 2M repairs may be performed only to meet an urgent operational commitment as directed by the operational commander.

SOURCE, MAINTENANCE, AND RECOVERABILITY (SM&R) CODES

The Allowance Parts List (APL) is a technical document prepared by the Navy for specific equipment/system support. This document lists the repair parts requirements for a ship having the exact equipment/component. To determine the availability of repair parts, the 2M technician must be familiar with these documents. SM&R codes, found in APLs, determine where repair parts can be obtained, who is authorized to make the repair, and at what maintenance level the item may be recovered or condemned.

Q5. What are the three levels of maintenance?

Q6. Maintenance performed by the user activity is what maintenance level?

TEST EQUIPMENT

Microelectronic developments have had a great impact on the test equipment, tools, and facilities necessary to maintain systems using this technology. This section discusses, in general terms, the importance of these developments.

Early electronic systems could be completely checked-out with general-purpose electronic test equipment (GPETE), such as multimeters, oscilloscopes, and signal generators. Using this equipment to individually test the microelectronics components in one of today's very complex electronic systems would be extremely difficult if not impossible. Therefore, improvements in system testing procedures have been necessary.

One such improvement in system testing is the design of a method that can test systems at various functional levels. This allows groups of components to be tested as a whole and reduces the time required to test components individually. One advantage of this method is that complete test plans can be written to provide the best sequencing of tests for wave shape or voltage outputs for each functional level. This method of testing has led to the development of special test sets, called AUTOMATED TEST EQUIPMENT (ATE). These test sets are capable of simulating actual operating conditions of the system being tested. Appropriate signal voltages are applied by the test set to the various functional levels of the system, and the output of each level is monitored. Testing sequences are prewritten and steps may be switched-in manually or automatically. The limits for each functional level are preprogrammed to give either a "go/no-go" indication or diagnose a fault to a component. A go/no-go indication means that a functional level either meets the test specifications (go) or fails to meet the specifications (no-go).

If a no-go indication is observed for a given function, the area of the system in which it occurs is then further tested. You can test the trouble area by using general purpose electronic test equipment and the troubleshooting manual for the system. General purpose electronic test equipment (GPETE) will be discussed later in this topic. (Effective fault isolation at this point depends on the experience of the technician and the quality of the troubleshooting manual.) After the fault is located, the defective part is then replaced or repaired, depending on the nature of the defect. At this stage, the defective part is usually a circuit card, a module, or a discrete part, such as a switch, relay, transistor, or resistor.

BUILT-IN TEST EQUIPMENT

One type of fault isolation that can be either on-line or off-line is BUILT-IN TEST EQUIPMENT (BITE). BITE is any device that is permanently mounted in the prime equipment (system); it is used only for testing the equipment or system in which it is installed either independently or in association with external test equipment. The specific types of BITE are too varied to discuss here, but may be as simple as a set of meters and switches or as complex as a computer-controlled diagnostic system.

ON-LINE TEST EQUIPMENT

Functional-level testing and modular design have been successfully applied to most electronic systems in use today; however, the trend toward increasing the number of subassemblies within a module by incorporating microelectronics will make this method of testing less and less effective.

The increased circuit density and packaging possible with microelectronic components makes troubleshooting and fault location difficult or, in some cases, impossible. The technician's efforts must be aided if timely repairs to microelectronic systems are to be achieved. These repairs are particularly significant when considered in the light of the very stringent availability requirements for today's systems. This dilemma has led to the present trend of developing both ON-LINE and OFF-LINE automatic test systems. The on-line systems are designed to continuously monitor performance and to automatically isolate faults to removable assemblies. Off-line systems automatically check removable assemblies and isolate faults to the component level.

Two on-line systems, the TEST EVALUATION AND MONITORING SYSTEM (TEAMS) and the CENTRALIZED AUTOMATIC TEST SYSTEM (CATS), are presently in production or under development by the Navy.

Test Evaluation and Monitoring System (TEAMS)

TEAMS is an on-line system that continuously monitors the performance of electronic systems and isolates faults to a removable assembly. This system is controlled by a computer using a test program on perforated or magnetic tape, cassettes, or disks. Displays are used to present the status of the equipment and to provide data with instructions for fault localization. Lights, usually an LED, are used to indicate which equipments are being tested and also which equipments are in an out-of-tolerance condition. A printer provides a read out copy of the test results. These results are used by maintenance personnel to isolate the fault in a removable assembly to a replaceable part.

Centralized Automatic Test System (CATS)

CATS is an on-line system that continuously monitors the performance of electronic systems, predicts system performance trends, and isolates faults to removable assemblies. CATS, however, is computer controlled and the instructions are preprogrammed in the computer memory. The status of the electronic system being monitored by CATS is presented in various forms. Information concerning a failed module is presented on a status- and fault-isolation indicator to alert the maintenance technician of the need for a replacement module. If equipment design does not permit module replacement, complete electrical schematics and fault-isolation procedures will be made available to the maintenance technician.

OFF-LINE TEST EQUIPMENT

The Navy has under development an advanced assembly tester designated Naval Electronics Laboratory Assembly Tester (NELAT). This tester is an off-line, general-purpose test system designed to check-out and isolate faults in electronic plug-in assemblies, modules, and printed circuit boards. Equipped with a complete range of instrumentation, the system allows testing to be accomplished automatically, semiautomatically, or manually. In the automatic mode, a complete range of stimuli generators and monitors are connected and switched by means of a microfilmed test program.

The NELAT incorporates modular electronic assemblies that will facilitate updating of the system. The system is designed for use aboard ship. When put into service, this tester will greatly improve the technician's capability in the checkout and fault isolation of microelectronic assemblies.

Another important system for off-line testing is the Versatile Avionic Shop Test System (VAST). VAST is used in the aviation community for fault isolation in aviation electronics (avionics) equipment on ships and shore commands with aircraft INTERMEDIATE MAINTENANCE DEPARTMENTS (AIMDs). It is an automatic, high-speed, computer controlled, general-purpose test set that will isolate faults to the component level.

GENERAL-PURPOSE ELECTRONIC TEST EQUIPMENT (GPETE)

When no automatic means of accomplishing fault isolation is available, general-purpose electronic test equipment and good troubleshooting procedures is used; however, such fault diagnosis should be attempted only by experienced technicians. Misuse of electrical probes and test equipment may permanently damage boards or microelectronic devices attached to them. The proximity of leads to one another and the effects of interconnecting the wiring make the testing of boards extremely difficult; these factors also make drift or current leakage measurements practically impossible.

Boards that have been conformally coated are difficult to probe because the coating is often too thick to penetrate for a good electrical contact. These boards must be removed for electrical probe testing. Many boards, however, are designed with test points that can be monitored either with special test sets or general-purpose test equipment. Another method of obtaining access to a greater number of test points is to use extender cards or cables. The use of extender cards or cables makes these test points easier to check.

Special care should be exercised when probing integrated circuits; they are easily damaged by excessive voltages or currents, and component leads may be physically damaged. Precautions concerning the use of test equipment for troubleshooting equipments containing integrated circuits are similar to those that should be observed when troubleshooting equipment containing semiconductor or other voltage and current-sensitive devices.

Voltage and resistance tests of resistors, transistors, inductors, and so forth, are usually effective in locating complete failures or defects that exhibit large changes from normal circuit characteristics; however, these methods are time-consuming and sometimes unsuccessful. The suspect device often must be desoldered, removed from the circuit, and then retested to verify the fault. If the defect is not verified, the device must be resoldered to the board again. If this procedure has to be repeated several times, or if the board is conformally coated, the defect may never be located. In fact, the circuit may be further damaged by the attempt to locate the fault. For these reasons, the device should never be desoldered until all possible in-circuit tests are performed and the defect verified.

Q7. List the three groups of test equipment used for fault isolation in 2M repair.

Q8. What test equipment continuously monitors electronic systems?

Q9. NELAT and VAST are examples of what type of test equipment?

REPAIR STATIONS

In addition to the requirements for special skills, the repair of 2M electronic circuits also requires special tools. Because these tools are delicate and expensive, they are distributed only to trained and certified 2M repair technicians.

2M repair stations are equipped with electrical and mechanical units, tools, and general repair materials. Such equipments are needed to make reliable repairs to miniature and microminiature component circuit boards.

Although most of the tools and equipments are common to both miniature and microminiature repair stations, several pieces of equipment are used solely with microminiature repair. Precision drill presses and stereoscopic-zoom microscopes are examples of microminiature repair equipment normally not found in a miniature repair station. A brief description of some of the tools and equipments and their uses will broaden your knowledge and understanding of 2M repair.

The 2M repair set consists of special electrical units, tools, and materials necessary to make high-reliability repairs to component circuitry. The basic repair set is made up of a repair station power unit, magnifier/light system, card holder, a high-intensity light, a Pana Vise, and a tool chest with specialized tools and materials. As mentioned previously, stations that have microminiature repair capabilities will include a stereoscopic-zoom microscope and precision drill press.

REPAIR STATION POWER UNIT

The repair station power unit is a standardized system that provides controlled soldering and desoldering of all types of solder joint configurations. The unit is shown in figure 2-1. Included in the control unit's capabilities are:

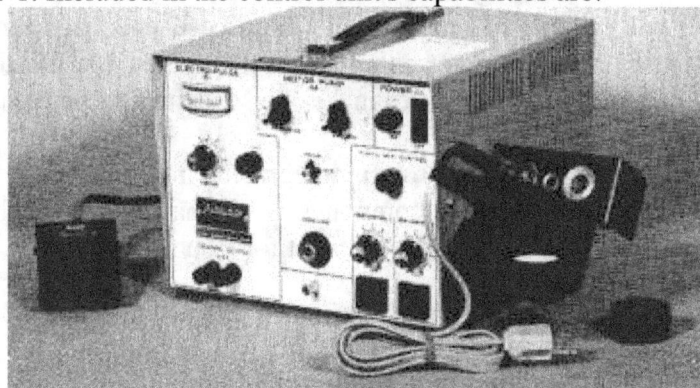

Figure 2-1.—Repair station power unit.

• "Spike free" power switching for attached electrical hand tools to eliminate damage to electrostatic discharge components.

• Abrading, milling, drilling, grinding, and cutting using a flexible shaft, rotary-drive machine. This allows the technician to remove conformal coatings, oxides, eyelets, rivets, damaged board material, and damaged platings from assemblies.

• Lap flow solder connections and thermal removal of conformal coatings.

• Resistive and conductive tweezer heating for connector soldering applications.

• Thermal wire stripping for removing polyvinyl chloride (PVC) and other synethetic wire coverings.

Power Source

The basic unit houses the power supply, power level indicator, motor control switch, hand tool temperature controls, air pressure and vacuum controls with quick connect fittings, positive ground terminal, the mechanical power-drive for the rotary-drive machine, and a vacuum/pressure pump. A two-position foot pedal, to the left of the power unit in the illustration, allows hand-free operation for all ancillary (additional) handpieces. The first detent on the pedal provides power to the voltage heating outputs. The second detent activates the motor drive or vacuum/pressure pump.

Handpieces

The handpieces used with the power unit are shown in figures 2-2 and 2-3. The lap flow handpiece, view (A) of figure 2-2, is used with the variable low-voltage power source. This handpiece allows removal of conformal coatings, release of sweat joints, and lap flow soldering capability. (Lap flow soldering will be discussed in topic 3.) The thermal wire stripper in view (B) is used to remove insulation from various sizes of wire easily and cleanly.

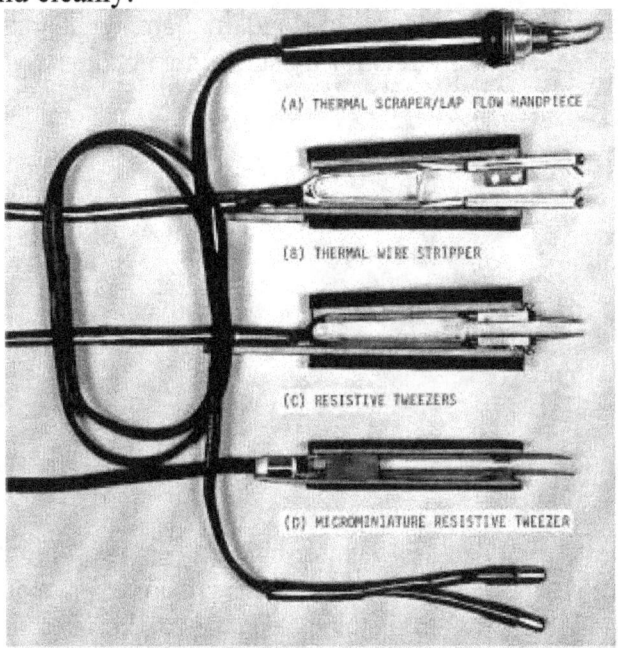

Figure 2-2.—Low voltage Handpiece.

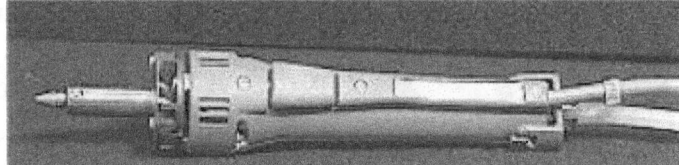

Figure 2-3.—Motorized solder extrator.

The resistive tweezers, shown in view (C), are used for soldering components. Two sizes [views (C) and (D)] are provided to meet the needs of the technician. Both the thermal stripper and the resistive tweezers are used with the low-voltage power supply.

The solder extractor, shown in view (A) of figure 2-3, is connected to the variable high-voltage outlet. This handpiece allows airflow application (at controlled temperatures) of a vacuum or pressure to the selected area. Five sizes of extractor tips are provided, as shown in view (B). You can determine the one to be used by matching the tip with the circuit pad and the component being desoldered.

Soldering Irons

A soldering iron is shown in figure 2-4. This is connected to the 115-volt ac variable outlet of the power unit. You control the temperature by adjusting the voltage. The iron has replaceable tips. Chosen for their long life and good heat conductivity, soldering iron tips are high quality with iron-clad over copper construction. The tip shape and size and the heat range used are determined by the area and mass to be soldered.

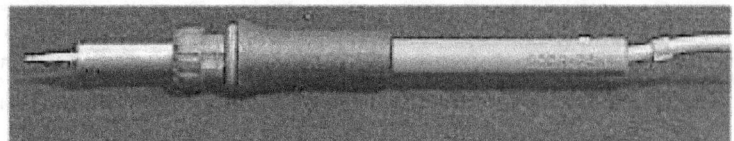
Figure 2-4.—Soldering iron.

ROTARY-DRIVE MACHINE

This variable-speed, rotary power drive adapts to standard diameter shank drill bits, ball mills, wheels, disks, brushes, and mandrels for most drilling and abrasive removal techniques (figure 2-5).

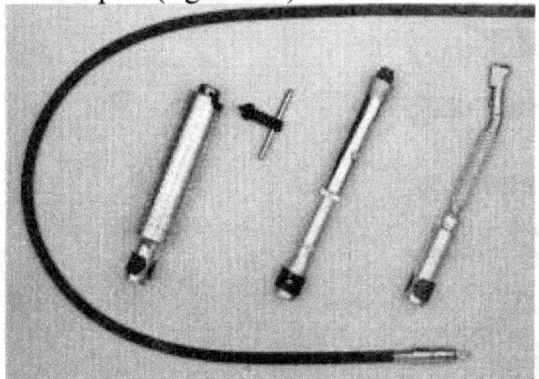
Figure 2-5.—Rotary-drive machine handpieces.

The accessories used with the rotary-drive tool are shown in views (A) through (F) of figure 2-6. Abrasive ball mills, wheels, discs, and brushes are either premounted on mandrels or can be mounted by the technician on the mandrels provided. These attachments are used for sanding and smoothing repaired areas, drilling holes, removing conformal coatings, and repairing burned or damaged areas. A chuck-equipped handpiece allows it to accept rotary tools with varying shank sizes.

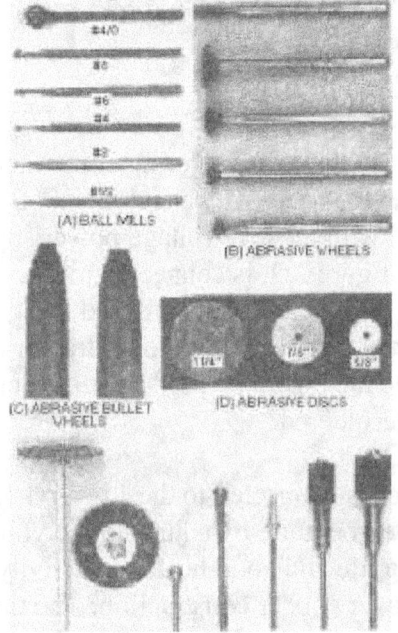
fEl DENTAL BRUSH (F) MANDRELS
Figure 2-6.—Rotary-drive machine accessories. BALL MILLS
CIRCUIT CARD HOLDER AND MAGNIFIER

The circuit card holder is an adjustable, rotatable holder for virtually any size circuit card. Figure 2-7 shows the circuit card holder [view (A)] and the magnifier unit [view (B)]. The magnifier unit provides magnification when detail provided by a microscope is not required. The special lens allows the technician to view a rectangular area of over 14 square inches with low distortion, fine resolution, and excellent depth of field. The magnifier unit, which includes high intensity lamps, adapts to the vertical shaft of the circuit card holder.

Figure 2-7.—Card holder and magnifier.

HIGH-INTENSITY LIGHT

The high-intensity light provides a variable, high-intensity, portable light source over the work area. The two flexible arms permit both front and back lighting of the workpiece and provide a balanced light that eliminates shadows (figure 2-8).

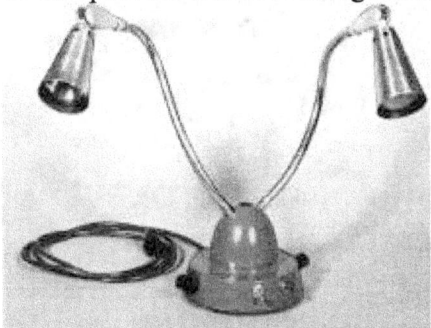

Figure 2-8.—High intensity lamp.

The high-intensity light uses 115-volt, 60-hertz input power. One brightness knob controls a flood-type bulb, and the other knob controls a spot-type bulb.

PANA VISE

This nylon-jawed, multiposition vise can rotate and tilt. With this flexibility the technician can achieve any compound angle for holding a workpiece during assembly, modification, or repair (figure 2-9).

Figure 2-9.—Pana Vise.

HAND TOOLS

Figure 2-10, views (A) through (C), shows some representative types of hand tools used in 2M repair procedures.

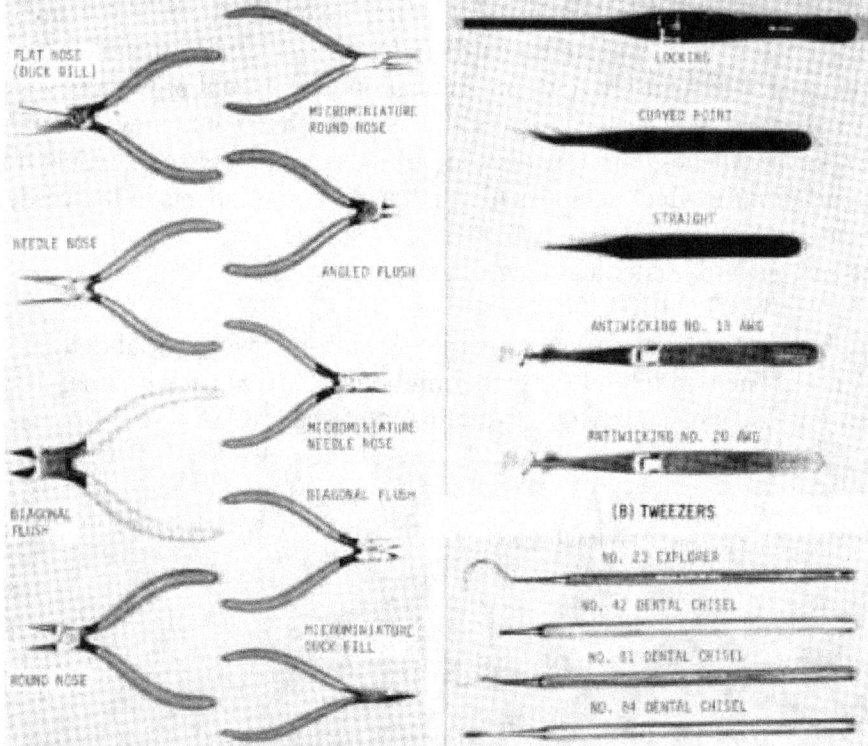

(A] PLIERS |C) DENTAL TOOLS

Figure 2-10.—Pliers, tweezers, and dental tools.

Pliers

In view (A), the figure shows the pliers preferred for 2M repair procedures. These precision pliers have a long and useful life if handled and cared for properly. The flush-cutting pliers are used to cut various sizes of wire and component leads. The needlenose, roundnose, and flatnose pliers are used for forming, looping, and bending wires and component leads. They are also used for gripping components and leads during removal or installation.

Figure 2-1 Oa.—Pliers.

Tweezers

View (B) shows tweezers contained in the 2M repair set. The top two pairs of tweezers are used to hold small components during installation and repair procedures. The other pairs are anti-wicking tweezers used to tin and solder stranded wire leads.

Dental Tools

View (C) shows some of the dental tools contained in the 2M repair set. They are used for picking, chipping, abrading, mixing, and smoothing various conformal coatings used on printed circuit boards and other general pcb repair techniques.

Figure 2-10c.—Dental tools.

Eyelet-Setting Tools

Among the repair procedures required of the 2M repair technician is the replacement of eyelets. Eyelets must sometimes be replaced because of the damage caused by incorrect repair procedures or complete failure of a printed circuit board. Figure 2-11 illustrates the tools used to replace these eyelets. Eyelets will be discussed in topic 3.

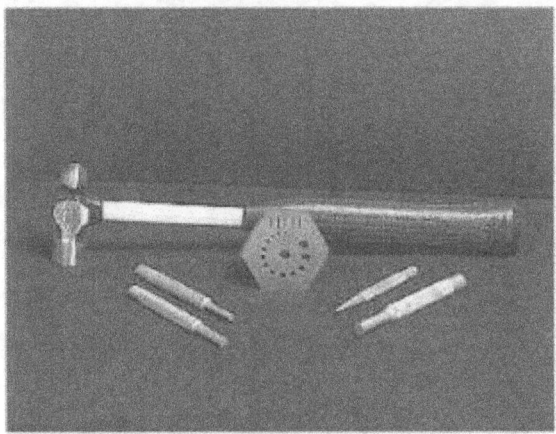

Figure 2-11.—Eyelet-setting tools.

MISCELLANEOUS TOOLS AND SUPPLIES

An assortment of some of the miscellaneous items used in 2M repair are shown in figure 2-12. A variety of brushes, files, scissors, thermal shunts, and consumables, such as solder wick, are included.

Even though all the items are not used in every repair procedure, it is extremely important that they be available for use should the need arise.

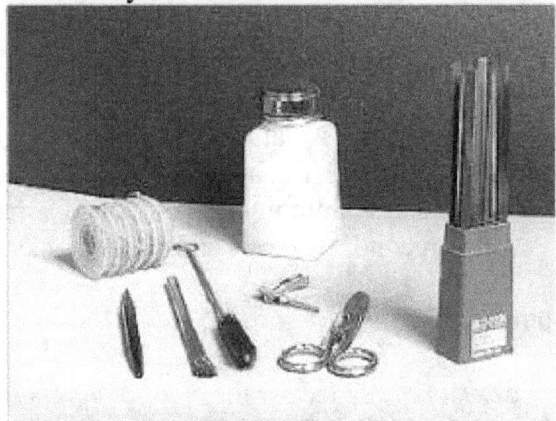

Figure 2-12.—Miscellaneous tools and supplies.

SAFETY EQUIPMENT

The nature of 2M repair requires items to be included in the tool kit for the personal safety of the technicians. The goggles and respirator illustrated in figure 2-13 have been approved for use by the technician. These should be worn at all times where dust, chips, fumes, and other hazardous substances are generated as a result of drilling, grinding, or other repair procedures.

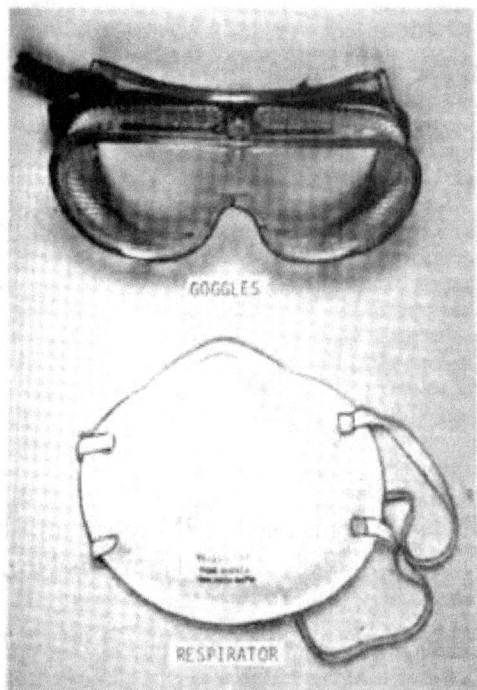

Figure 2-13.—Safety equipment.

STEREOSCOPIC-ZOOM MICROSCOPE

The stereoscopic-zoom microscope provides a versatile optical viewing system. This viewing system is used in the fault detection, fault isolation, and repair of complex microminiature circuit boards and components. Figure 2-14 shows the microscope mounted on an adjustable stand. The microscope has a minimum of 3.5X and a maximum of 30X magnification to detect hairline cracks in conductor runs and stress cracks in solder joints.

Figure 2-14.—Stereoscopic zoom microscope.

TOOL CHEST

The tool chest (not shown), provides storage space for the electronic repair hand tools, dental tools, abrasive wheels, solder and solder wicks, eyelets, abrasive disks, ball mills, various burrs, and other consumables used with the repair procedures. The chest is portable, lockable, and has variously sized drawers for convenience.

REPLACEMENT PARTS

Replacement parts are provided with the 2M repair set to ensure the technician has the capability to maintain the equipment properly. Actual preventive and corrective maintenance procedures, as well as data on additional spare parts and ordering information, are found in the technical manual for the 2M repair set equipment.

REPAIR STATION FACILITIES

To be effective, 2M electronic component repair must be performed under proper environmental conditions. Repair facility requirements, whether afloat or ashore, include adequate lighting, ventilation, noise considerations, work surface area, ESD (electrostatic discharge) protection, and adequate power availability. The recommended environmental conditions are discussed below. With the exception of requirements imposed by the Naval Environmental Health Center and other authorities for ship and shore work conditions, each activity tailors the requirements to meet local needs.

LIGHTING

The recommended lighting for a work surface is 100 footcandles from a direct lighting source. Light-colored overheads and bulkheads and off-white or pastel workbench tops are used to complement the lighting provided.

VENTILATION

Fumes from burning flux, coating materials, grinding dust, and cleaning solvents require adequate ventilation. The use of toxic, flammable substances, solvents, and coating compounds requires a duct system that vents gasses and vapors. This type of system must be used to prevent contamination often

found in closed ventilation systems. This need is particularly important aboard ship. Vented hoods, ducts, or installations that are vented outside generally meet the minimum standards set by the Naval Environmental Health Center.

NOISE CONSIDERATIONS

Noise in the work area during normal work periods must be no greater than the acceptable level approved for each activity involved. Because the work is tedious and tiring, noise levels should be as low as possible. Ear protectors are required to be worn when a noise level exceeds 85 dB. Ear protectors should also be worn anytime the technician feels distracted by, or uncomfortable with, the noise level.

WORK SURFACE AREA

Work stations should have a minimum work surface of at least 60-inches wide and 30-inches deep. Standard Navy desks are excellent for this purpose. Standard shipboard workbenches are acceptable; however, off-white or pastel-colored heat-resistant tops should be installed on the workbenches. Chairs should be the type with backs and without arms. They should be comfortably padded and of the proper height to match the work surface height. Drawers or other suitable tool storage areas are usually provided.

ELECTROSTATIC DISCHARGE SENSITIVE DEVICE (ESDS) CAPABILITY

A 2M work station should be capable of becoming a static-free work station. This is specified in the Department of Defense Standard, Electrostatic DISCHARGE Control Program for Protection of Electrical and Electronic Parts, Assemblies, and Equipment. ESD will be discussed in greater detail in topic 3.

POWER REQUIREMENTS

No special power source or equipment mounting is required. The 2M repair equipment operates on 115-volt, 60-hertz power. A 15-ampere circuit is sufficient and six individual power receptacles should be available.

HIGH-RELIABILITY SOLDERING

The most common types of miniature and microminiature repair involve the removal and replacement of circuit components. The key to these repairs is a firm knowledge of solder and high-reliability soldering techniques.

Solder is a metal alloy used to join two or more metals with a metallic bond. The bonding occurs when molten solder dissolves a small amount of the metals and then cools to form a solid connection. The solder most commonly used in electronic assemblies is an alloy of tin and lead. Tin-lead alloys are identified by their percentage in the solder; the tin content is given first. Solder marked 60/40 is an alloy of 60 percent tin and 40 percent lead. The two most common alloys used in electronics are 60/40 and 63/37.

The melting temperature of tin-lead solder varies depending on the percentage of each metal. Lead melts at a temperature of 621 degrees Fahrenheit, and tin melts at 450 degrees Fahrenheit. Combinations of the two metals melt into a liquid at different temperatures. The 63/37 combination melts into a liquid at 361 degrees Fahrenheit. At this temperature, the alloy changes from a solid directly to a liquid with no plastic or semiliquid state. An alloy with such a sharp changing point is called a EUTECTIC ALLOY.

As the percentages of tin and lead are varied, the melting temperature increases. Alloy of 60/40 melts at 370 degrees Fahrenheit, and alloy of 70/30 melts at approximately 380 degrees Fahrenheit. Alloys,

other than eutectic, go through a plastic or semiliquid state in their heating and cooling stages. Solder joints that are disturbed (moved) during the plastic state will result in damaged connections. For this reason, 63/37 solder is the best alloy for electronic work. Solder with 60/40 alloy is also acceptable, but it goes into a plastic state between 361 and 370 degrees Fahrenheit. When soldering joints with 60/40 alloy, you must exercise extreme care to prevent movement of the component during cooling.

USE OF FLUX IN SOLDER BONDING

Reliable solder connections can only be accomplished with clean surfaces. Using solvents and abrasives to clean the surfaces to be soldered is essential if you are to achieve good solder connections. In almost all cases, however, this cleaning process is insufficient because oxides form rapidly on heated metal surfaces. The rapid formation of oxides creates a nonmetallic film that prevents solder from contacting the metal. Good metal-to-metal contact must be obtained before good soldering joints may take place. Flux removes these surface oxides from metals to be soldered and keeps them removed during the soldering operation. Flux chemically breaks down surface oxides and causes the oxide film to loosen and break free from the metals being soldered.

Soldering fluxes are divided into three classifications or groups: CHLORIDE FLUX (commonly called ACID), ORGANIC FLUX, and ROSIN FLUX. Each flux has characteristics specific to its own group. Chloride fluxes are the most active of the three groups. They are effective on all common metals except aluminum and magnesium. Chloride fluxes, however, are NOT suitable for electronic soldering because they are highly corrosive, electrically conductive, and are difficult to remove from the soldered joint.

Organic fluxes are nearly as active as chloride fluxes, yet are less corrosive and easier to remove than chloride fluxes. Also, these fluxes are NOT satisfactory for electronic soldering because they must be removed completely to prevent corrosion.

Rosin fluxes ARE ideally suited to electronic soldering because of their molecular structure. The most common flux used in electronic soldering is a solution of pure rosin dissolved in suitable solvent. This solution works well with the tin- or solder-dipped metals commonly used for wires, lugs, and connectors. While inert at normal temperatures, rosin fluxes break down and become highly active at soldering temperatures. In addition, rosin is nonconductive.

Most electronic solder, in wire form, is made with one or more cores of rosin flux. When the joint or connection is heated and the wire solder is applied to the joint (not the iron), the flux flows onto the surface of the joint and removes the oxide. This process aids the wetting action of the solder. With enough heat the solder flows and replaces the flux. Insufficient heat results in a poor connection because the solder does not replace the flux.

Q10. Stereoscopic-zoom microscopes and precision drill presses are normally associated with what type of repair station?

Qll. Solder used in electronic repair is normally an alloy of what two elements?

Q12. In soldering, what alloy changes directly from a solid state to a liquid state?

Q13. Flux aids in soldering by removing what from surfaces to be soldered?

Q14. What type(s) of flux should never be used on electronic equipment?

SUMMARY

This topic has presented information on the Miniature and Microminiature 2M Repair Program and high-reliability soldering. The information that follows summarizes the important points of this topic.

The MINIATURE/MICROMINIATURE (2M) REPAIR PROGRAM provides training, tools and equipment, and certification for 2M repair personnel.

CERTIFICATION of technicians ensures the capability of high-quality, high-reliability repairs.

The three SM&R codes for maintenance of electronic devices are: DEPOT (D), INTERMEDIATE (I), and ORGANIZATIONAL (O).

SM&R CODE D MAINTENANCE is characterized by extensive facilities and highly trained personnel. Code D activities are capable of the most complex type repairs.

CODE I activities provide direct support for user activities. This includes calibration, repair, and emergency manufacture of nonavailable parts.

CODE O maintenance is the responsibility of the user activity. It includes preventive maintenance and minor repairs.

ON-LINE TEST EQUIPMENT continuously monitors system performance and isolates faults to removable assemblies.

OFF-LINE TEST EQUIPMENT evaluates removable assemblies outside of the equipment and isolates faults to the component level.

FAULT ISOLATION USING GENERAL-PURPOSE ELECTRONIC TEST EQUIPMENT (GPETE) should only be attempted by experienced technicians.

2M REPAIR STATIONS are equipped according to the level of repairs to be accomplished.

ALLOYS, such as solder, which change directly from a solid state to a liquid are called eutectic alloys.

SOLDER with a tin/lead ratio of 63/37 is preferred for electronic work. A ratio of 60/40 is also acceptable.

ROSIN or RESIN FLUXES are the only fluxes to be used in electronic work.

ANSWERS TO QUESTIONS Ql. THROUGH Q14.

Al. Chief of Naval Operations (CNO).

A2. Naval Sea Systems Command (NA VSEASYSCOM) and Naval Air Systems Command (NAVAIRSYSCOM).

A3. Microminiature component repair.

A4. Microminiature repair technician.

A5. Depot, Intermediate, and Organizational.

A6. Organizational.

A7. On-line, off-line, and General Purpose Electronic Test Equipment (GPETE).

A8. On-line.

A9. Off-line.

A10. Microminiature repair station.

All. Tin and lead.

A12. Eutectic.

A13. Oxides.

A14. Chloride or (acid) and organic.

CHAPTER 3

MINIATURE AND MICROMINIATURE REPAIR

PROCEDURES

LEARNING OBJECTIVES

Upon completion of this topic, the student will be able to:

1. Explain the purpose of conformal coatings and the methods used for removal and replacement of these coatings.

2. Explain the methods and practices for the removal and replacement of discrete components on printed circuit boards.

3. Identify types of damage to printed circuit boards, and describe the repair procedures for each type of repair.

4. Describe the removal and replacement of the dual-in-line integrated circuit.

5. Describe the removal and replacement of the TO-5 integrated circuit.

6. Describe the removal and replacement of the flat-pack integrated circuit.

7. Describe the types of damage to which many microelectronic components are susceptible and methods of preventing damage.

8. Explain safety precautions as they relate to 2M repair.

INTRODUCTION

As you progress in your training as a technician, you will find that the skill and knowledge levels required to maintain electronic systems become more demanding. The increased use of miniature and microminiature electronic circuits, circuit complexity, and new manufacturing techniques will make your job more challenging. To maintain and repair equipment effectively, you will have to duplicate with limited facilities what was accomplished in the factory with extensive facilities. Printed circuit boards that were manufactured completely by machine will have to be repaired by hand.

To meet the needs for repairing the full range of electronic equipment, you must be properly trained. You must be capable of performing high-quality, reliable repairs to the latest circuitry.

MINIATURE AND MICROMINIATURE ELECTRONIC REPAIR

PROCEDURES

As mentioned at the beginning of topic 2, 2M repair personnel must undergo specialized training. They are trained for a particular level of repair and must be certified at that level. Also, recertification is required to ensure the continued high-quality repair ability of these technicians.

CAUTION

THIS SECTION IS NOT, IN ANY WAY, TO BE USED BY YOU AS AUTHORIZATION TO ATTEMPT THESE TYPES OF REPAIRS WITHOUT OFFICIAL 2M CERTIFICATION.

In the following sections, you will study the general procedures used in the repair, removal, and replacement of specific types of electronic components. By studying these procedures, you will become familiar with some of the more common types of repair work. Before repair work can be performed on a miniature or microminiature assembly, the technician must consider the type of specialized coating that usually covers the assembly. These coatings are referred to as CONFORMAL COATINGS.

CONFORMAL COATINGS

Conformal coatings are protective material applied to electronic assemblies to prevent damage from corrosion, moisture, and stress. These coatings include epoxy, parylene, silicone, polyurethane, varnish, and lacquer. Coatings are applied in a liquid form; when dry, they exhibit characteristics that improve reliability. These characteristics are:

- Heat conductivity to carry heat away from components
- Hardness and strength to support and protect components
- Low moisture absorption
- Electrical insulation

Conformal Coating Removal

Because of the characteristics that conformal coatings exhibit, they must be removed before any work can be done on printed circuit boards. The coating must be removed from all lead and pad/eyelet areas of the component. It should also be removed to or below the widest point of the component body. Complete removal of the coating from the board is not done.

Methods of coating removal are thermal, mechanical, and chemical. The method of removal depends on the type of coating used. Table 3-1 shows suggested methods of removal of some types. Note that most of the methods are variations of mechanical removal.

Table 3-1.—Conformal Coating Removal Techniques

TYPES

CONFORMAL COATING (LISTED IN DESCENDING ORDER OF HARDNESS)

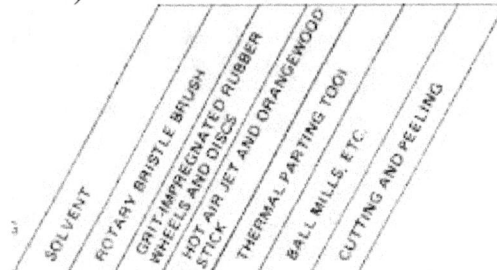

* DENOTED METHOD THAT POSITIVELY IDENTIFIES TYPE OF CONFORMAL COATING. A SPECI FIC TYPES OF COATING COMPOSITIONS ONLY. 6 FOR THICK COATINGS ONLY (.025 AND THICKER!. C FOR THIN COATINGS ONLY.

D DO NOT ATTEMPT TO GRIND TO BOARD SURFACE WITH THIS METHOD. E ORGANIC SOLVENTS ARE BEST.

F USE MODIFIED DRILL BIT FOR TIP (SHAPED TO BEVELED EDGE) CAUTION: USE WITH CAFE, APPROXIMATELY 5D0 0 NEEDED AT TIP.

NOTE' THE PREFEHRED OHDER FOR APPLYING INDIVIDUAL REMOVAL TECHNIQUES TO SPECIFIC COATINGS IS NUMERICALLY INDICATED. THESE REMOVAL TECHNIQUES ARE LISTED IN ASCENDING OHDER OF THEIR DAMAGEABILITY TO THE MODULE UNDER REPAIR. ANY OF THE METHODS LISTED MAY CAUSE DAMAGE IF NOT USED WITH CARE,

ALWAYS TRY THE METHOD WHICH CAUSES THE LEAST AMOUNT OF DAMAGE FIRST. CONSIDER THE POSSIBILITY OF HEAT OR VIBRATION SENSITIVITY OF COMPONENTS. (VIBRATION ESPECIALLY WILL AFFECT ALL AREAS OF THE WORK-PIECE.)

The coating material can best be identified through proper documentation; for example, technical manuals and engineering drawings. If this information is not available, the experienced technician can usually determine the type of material by testing the, hardness, transparency, thickness, and solvent solubility of the coating. The thermal (heat) properties may also be tested to determine the ease of removal of the coating by heat. The methods of removal discussed here describe the basic concept, but not the step-by-step "how to" procedures.

THERMAL REMOVAL.—Thermal removal consists of using controlled heat through specially shaped tips attached to a handpiece. Soldering irons should never be used for coating removal because the high temperatures will cause the coatings to char, possibly damaging the board materials. Modified tips or cutting blades heated by soldering irons also are not used; they may not have proper heat capacity or allow the hand control necessary for effective removal. Also, the thin plating of the circuit may be damaged by scraping.

The thermal parting tool, used with the variable power supply, has interchangeable tips, as shown in figure 3-1, that allow for efficient coating removal. These thin, blade-like instruments act as heat generators and will maintain the heat levels necessary to accomplish the work. Tips can be changed easily to suit the configuration of the workpiece. These tips cool quickly after removal of power because their small thermal mass and special alloy material easily give up residual heat.

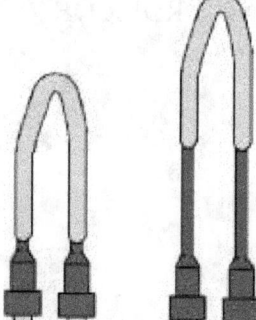

Figure 3-1.—Thermal parting tips.

The softening or breakdown point of different coatings vary, which is a concern when you are using this method. Ideally, the softening, point is below the solder melting temperature. However, when the softening point is equal to or above the solder melting point, you must take care in applying heat at the solder joint or in component areas. The work must be performed rapidly to limit the heating of the area involved and to prevent damage to the board and other components.

HOT-AIR JET REMOVAL.—In principle, the hot-air jet method of coating removal uses controlled, temperature-regulated air to soften or break down the coating, as shown in figure 3-2. By controlling the temperature, flow rate, and shape of the jet, you may remove coatings from almost any workpiece configuration without causing any damage. When you use the hot-air jet, you do not allow it to

physically contact the workpiece surface. Delicate work handled in this manner permits you to observe the removal process.

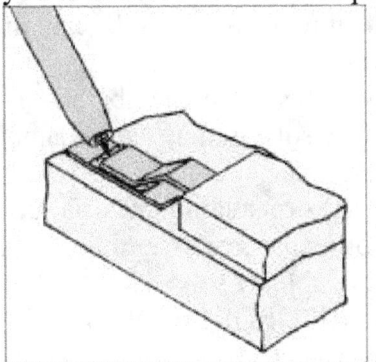

Figure 3-2.—Hot air jet conformal coating removal.

POWER-TOOL REMOVAL DESCRIPTION.—Power-tool removal is the use of abrasive grinding or cutting to mechanically remove coatings. Abrasive grinding/rubbing techniques are effective on thin coatings (less than 0.025 inch) while abrasive cutting methods are effective on coatings greater than 0.025 inch. This method permits consistent and precise removal of coatings without mechanical damage or dangerous heating to electronic components. A variable-speed mechanical drive handpiece permits fingertip-control and proper speed and torque to ease the handling of gum-type coatings. A variety of rotary abrasive materials and cutting tools is required for removal of the various coating types. These specially designed tools include BALL MILLS, BURRS, and ROTARY BRUSHES.

The ball mill design places the most efficient cutting area on the side of the ball rather than at the end. Different mill sizes are used to enter small areas where thick coatings need to be removed (ROUTED). Rubberized abrasives of the proper grade and grit are ideally suited for removing thin, hard coatings from flat surfaces; soft coatings adhere to and coat the abrasive causing it to become ineffective. Rotary bristle brushes work better than rubberized abrasives on contoured or irregular surfaces, such as soldered connections, because the bristles conform to surface irregularities. Ball mill routing and abrasion removal are shown in figure 3-3.

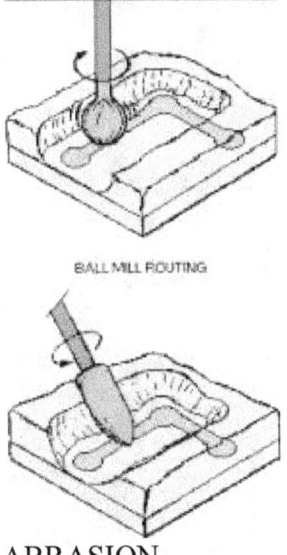

BALL MILL ROUTING

ABRASION

Figure 3-3.—Rotary tool conformal coating removal.

CUT AND PEEL.—Silicone coatings (also referred to as RTV) can easily be removed by cutting and peeling. As with all mechanical removal methods, care must be taken to prevent damage to either components or boards.

CHEMICAL REMOVAL.—Chemical removal uses solvents to break down the coatings. General application is not recommended as the solvent may cause damage to the boards by dissolving the adhesive materials that bond the circuits to the boards. These solvents may also dissolve the POTTING COMPOUNDS (insulating material that completely seals a component or assembly) used on other parts or assemblies. Only thin acrylic coatings (less than 0.025 inch) are readily removable by solvents. Mild solvents, such as ISOPROPYL ALCOHOL, XYLENE, or TRICHLOROETHANE, may be used to remove soluble coatings on a spot basis.

Evaluations show that many tool and technique combinations have proven to be reliable and effective in coating removal; no single method is the best in all situations. When the technician is determining the best method of coating removal to use, the first consideration is the effect that it will have on the equipment.

Conformal Coating Replacement

Once the required repairs have been completed the conformal coating must be replaced. To ensure the same protective characteristics, you should use the same type of replacement coating as that removed.

Conformal coating application techniques vary widely. These techniques depend on material type, required thickness of application, and the effect of environmental conditions on curing. These procedures cannot be effectively discussed here.

Ql. What material is applied to electronic assemblies to prevent damage from corrosion, moisture, and stress?

Q2. What three methods are used to remove protective material? Q3. What chemicals are used to remove protective material?

Q4. Abrasion, cutting, and peeling are examples of what type ofprotective material removal?

Q5. Why should the coating material be replaced once the required repair has been completed?

REMOVAL AND REPLACEMENT OF DISCRETE COMPONENTS

To properly perform the required repair, the 2M technician must be knowledgeable of the techniques used by manufacturers in the production of electronic assemblies. The techniques, materials, and types of components determine the repair procedures used.

Interconnections and Assemblies

Assemblies may range from simple, single-sided boards with standard-sized components to double-sided or multilayered boards with miniature and microminiature components. The variations in component lead termination and mounting techniques used by manufacturers present the technician with a complex task. For example, the 2M technician is concerned about the type of solder joints on the module. To determine the solder joint type, the technician must consider the board circuitry, hole reinforcement, and lead termination style.

Recall the discussion from topic 1 on printed circuit board construction and the types of interconnections used. Single-sided and some double-sided boards have UNSUPPORTED HOLES where component leads are soldered to the pad. The clearance-hole method is also an interconnection with no hole support. SUPPORTED HOLES are those that have metallic reinforcement along the hole walls.

In addition to the plated-through hole you studied earlier, EYELETS, shown in figure 3-4, view (A), view (B), and view (C), are also used in both manufacturing and repair. These hole-reinforcing devices are usually made of pure copper, but are often plated with gold, tin, or a tin-lead alloy. The copper-based eyelet is pliable; when set, it reduces the possibility of circuit board damage. Eyelets may be inserted into single-sided or double-sided boards and are of three different types - ROLL SET, FUNNEL SET, and FLAT SET. All three are types referred to as INTERFACIAL CONNECTIONS. Interfacial connections identify the procedure of connecting circuitry on one side of a board with the circuitry on the other side.

(A) ROLL SET Figure 3-4A.—Eyelets (interfacial connections). ROLL SET

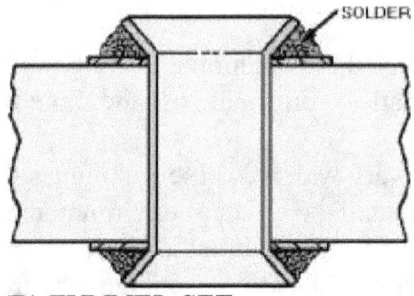

(B) FUNNEL SET

Figure 3-4B.—Eyelets (interfacial connections). FUNNEL SET

(C) FLAT SET

Figure 3-4C.—Eyelets (interfacial connections). FLAT SET

As you can see, the flat-set eyelet actually provides reinforcement for the pads on both sides of the circuit board and reinforces the hole itself. The design of the roll-set eyelet (which may trap gasses, flux, or other contaminants, and obscures view of the finished solder flow) is not acceptable as a repair technique. The funnel-set eyelet does not provide as much pad reinforcement as the other types. However, it provides better "outgassing" of flux, moisture, or solvents from the space between the eyelet and the hole wall. It also provides a better view of the finished solder connection than the roll-set eyelet.

Lead Terminations

The finished circuit board consists of conductive paths, pads, and drilled holes with components and/or wires assembled directly to it. Leads and wires may terminate in three ways: (1) through the hole in the board, (2) above the surface of the board, or (3) on the surface of the board.

THROUGH-HOLE TERMINATION.—This style provides extra support for the circuit pads, the hole, and the lead by a continuous solder connection from one side of the circuit board to the other. Three basic variations of through-hole termination are the CLINCHED LEAD (two types), STRAIGHT-THROUGH LEAD, and OFFSET PAD.

Clinched Lead.—The clinched-lead termination is usually used with unsupported holes, but is found with supported holes as well. Both clinched-lead types, FULLY CLINCHED and SEMICLINCHED (figure 3-5), provide component stability. Like the fully clinched lead, the semi-clinched lead also provides stability during assembly. However, this termination can be easily straightened to allow removal of the solder joint should rework or repair be required. Note that the fully clinched lead is bent 90 degrees while the semiclinched lead is bent 45 degrees.

fV /// A \

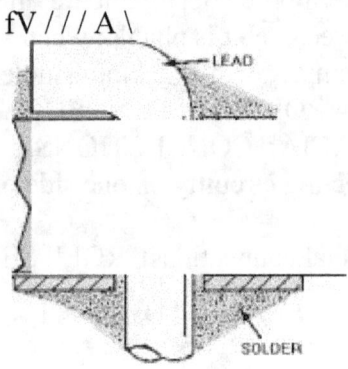

(A) FULLY CLINCHED

Figure 3-5A.—Clinched leads. FULLY CLINCHED.

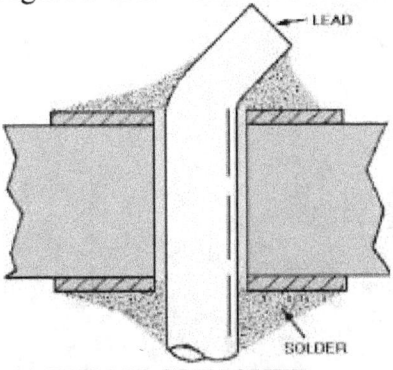

(A) FULLY CLINCHED

Figure 3-5B.—Clinched leads. SEMICLINCHED

Straight-Through Lead.—Straight-through terminations (figure 3-6) are used by manufacturers when the termination stability is not a prime consideration. This termination type may also be used with unsupported holes. The through-hole termination provides a better, solder-joint contact area and more solder support; the solder runs from the component side to the conductor. The straight-through termination is the easiest to remove and rework.

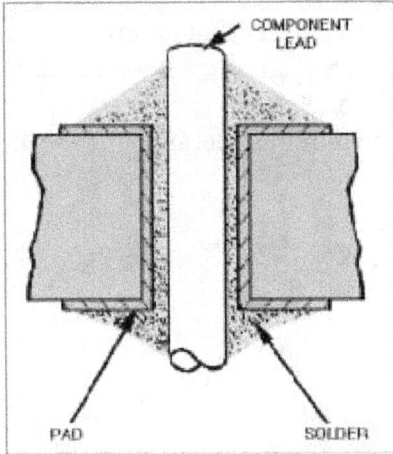

Figure 3-6.—Straight-through termination.

The Offset-Pad Termination.—This termination, shown in view (A) of figure 3-7, is a variation of clinch-lead termination. The pad is set off from the centerline of the hole. The lead clinch is also offset from the hole centerline so that it may contact the pad [view (B)].

COMPONENT LEAD

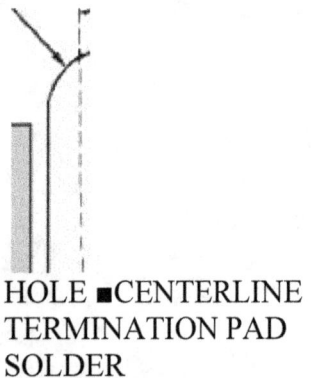

HOLE ■CENTERLINE
TERMINATION PAD
SOLDER
/ X/J///f777777.
Figure 3-7A.—Offset pad termination SIDE VIEW.

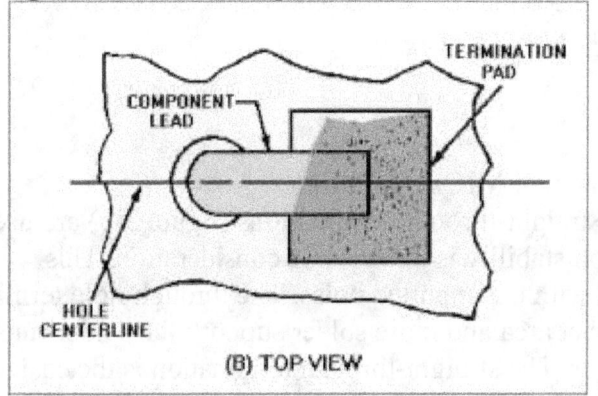

Figure 3-7B.—Offset pad termination TOP VIEW

ABOVE-THE-BOARD TERMINATION.—Above-the-board termination is accomplished through the use of terminals or posts. Terminals are used for a variety of reasons. The type of terminal depends on its use. Although many configurations are used, all terminals fall into one of the five categories covered in this section [figure 3-8, views (A) through (E)].

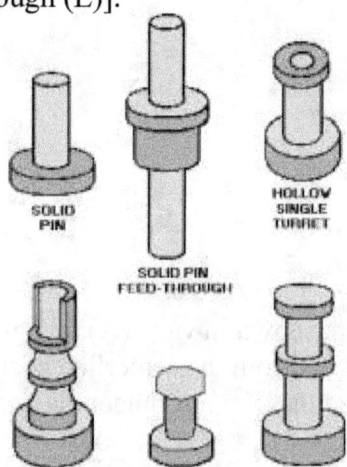

HOLLOV TURRET. SINGLE DOUBLE SPLIT POST TURRET TURRET
(A) PIN AND TERMINALS
Figure 3-8A.—Terminals. PIN AND TERMINALS.

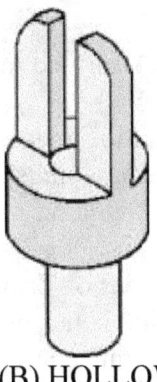

(B) HOLLOW BIFURCATED
Figure 3-8B.—Terminals. HOLLOW.

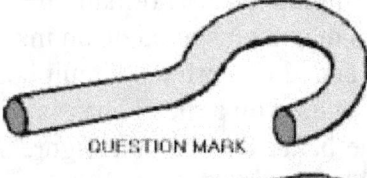

QUESTION MARK

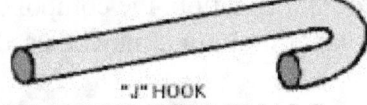

"J" HOOK

(C) HOOK TERMINALS
Figure 3-8C—Terminals. HOOK TERMINALS.

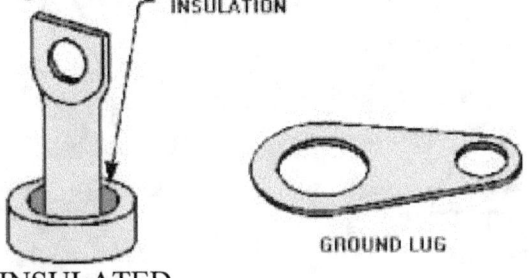

INSULATION

GROUND LUG

INSULATED
(D) PIERCED TERMINALS
Figure 3-8D.—Terminals. PIERCED TERMINALS.

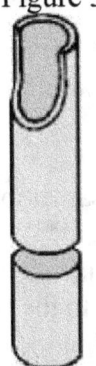

(E) SOLDER CUP
Figure 3-8E.—Terminals. SOLDER CUP

• PIN TERMINALS AND TURRET TERMINALS [view (A)] are single-post terminals, either insulated or uninsulated, solid or hollow, stud or feed-through. Stud terminals protrude from one side of a board; feed-throughs protrude from both sides.

    • BIFURCATED OR FORK TERMINALS [view (B)] are solid or hollow double-post terminals.

• HOOK TERMINALS [view (C)] are made of cylindrical stock formed in the shape of a hook or question mark.

• PERFORATED OR PIERCED TERMINALS [view (D)] describe a class of terminals that uses a hole pierced in flat metal for termination (e.g., terminal lugs).

• SOLDER CUP TERMINALS [view (E)] are a common type found on connectors.

Turret and bifurcated terminals are used for interfacial connections on printed circuit boards, terminal points for point-to-point wiring, mounting components, and as tie points for interconnecting wiring. Hook terminals are used to provide connection points on sealed devices and terminal boards.

Terminals used for wire or component lead terminations are normally made of brass with a solderable coating. Uninsulated terminals may be installed on an insulating substrate to form a terminal board. They may also be added to a printed circuit board or installed on a metal chassis. Insulated terminals are installed on a metal chassis.

ON-THE-BOARD TERMINATION.—On-the-board termination (figure 3-9) is also called LAP FLOW termination. In a lap flow solder termination, the component lead does not pass through the circuit board. This form of planar mounting may be used with both round and flat leads.

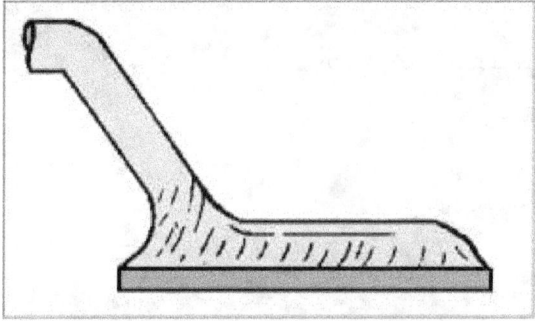

Figure 3-9.—On-the-board termination.

Q6. What term is used to identify the procedure of connecting one side of a circuit board with the other?

Q7. Name two types of through-hole termination.

Q8. Turret, bifurcated, and hook terminals are used for what type of termination?

Q9. When a lead is soldered to a pad without passing through the board, it is known as what type of termination?

Component Desoldering

Most of the damage in printed circuit board repair occurs during disassembly or component removal. More specifically, much of this damage occurs during the desoldering process. To remove components for repair or replacement, the technician must first determine the type of joint that is used to connect the component to the board. The technician may then determine the most effective method for desoldering these connections.

Three generally accepted methods of solder connection removal involve the use of SOLDER WICK, a MANUALLY CONTROLLED VACUUM PLUNGER, or a motorized solder extractor using CONTINUOUS VACUUM AND/OR PRESSURE. Of all the extraction methods currently in use, continuous vacuum is the most versatile and reliable. Desoldering becomes a routine operation and the quantity and quality of desoldering work increases with the use of this technique.

SOLDER WICKING.—IN this technique, finely stranded copper wire or braiding (wick) is saturated with liquid flux. Most commercial wick is impregnated with flux; the liquid flux adds to the effectiveness of the heat transfer and should be used whenever possible. The wick is then applied to a solder joint between the solder and a heated soldering iron tip, as shown in figure 3-10. The combination of heat, molten solder, and air spaced in the wick creates a capillary action and causes the solder to be drawn into the wick.

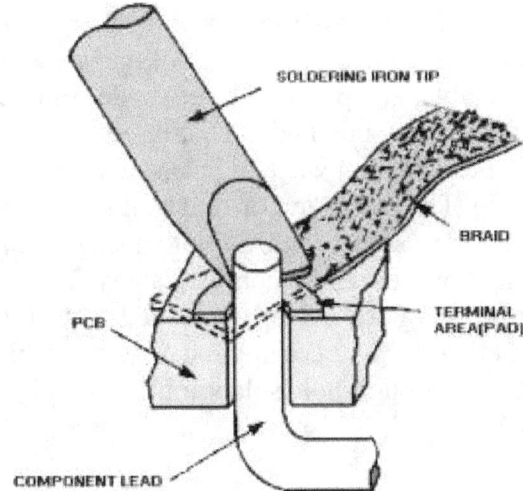

Figure 3-10.—Solder wicking.

This method should be used to remove surface joints only, such as those found on single-sided and double-sided boards without plated-through holes or eyelets. It can also remove excessive solder from flat surfaces and terminals. The reason is that the capillary action of the wicking is not strong enough to overcome the surface tension of the molten solder or the capillary action of the hole.

MANUALLY CONTROLLED VACUUM PLUNGER.—The second method of removing solder involves a manually controlled and operated, one-shot vacuum source. This vacuum source uses a plunger mechanism with a heat resistant orifice. The vacuum is applied through this orifice. Figure 3-11 shows the latest approved, manual-type desoldering tool. This technique involves melting the solder joint and inserting the solder-extractor tip into the molten solder over the soldering iron tip. The plunger is then released, creating a short pulse of vacuum to remove the molten solder. Although this method offers a positive vacuum rather than the capillary force of the wicking method, it still has limited application. This method will not remove 100 percent of the solder and may cause circuit pad lifting because of the extremely high vacuum generated and the jarring caused by the plunger action.

CARBON METALIZED LOADED BODY TIP /

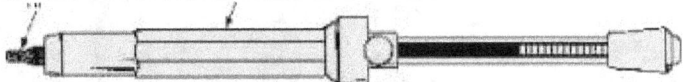

Figure 3-11.—Manual desoldering tool.

Because 100 percent of the solder cannot be removed, the extraction method is not usually successful with the plated-through solder joint. The component lead in a plated-through hole joint usually rests against the side wall of the hole. Even though most of the molten solder is removed by a vacuum, the small amount of solder left between the lead and side walls causes a SWEAT JOINT to form. A sweat joint is a paper-thin solder joint formed by a minute amount of solder remaining on the conductor lead surfaces.

MOTORIZED VACUUM/PRESSURE METHOD.—The most effective method for solder joint removal is motorized vacuum extraction. The solder extractor unit, described in topic 2, is used for this type of extraction. This method provides controlled combinations of heat and pressure or vacuum for solder removal. The motorized vacuum is controlled by a foot switch and differs from the manual vacuum in that it provides a continuous vacuum. The solder extraction device is a coaxial, in-line instrument similar to a small soldering iron. The device consists of a hollow-tipped heating element, transfer tube, and collecting chamber (in the handle) that collects and solidifies the waste solder. This unit is easily maneuvered, fully controllable, and provides three modes of operation (figure 3-12): (1) heat and vacuum (2) heat and pressure, and (3) hot-air jet. Some power source models provide variable control for pressure and vacuum levels as well as temperature control for the heated tubular tip. The extraction tip and heat source are combined in one tool. Continuous vacuum allows solder removal with a single heat application. Since the slim heating element allows access to confined areas, the technician is protected from contact with the hot, glass, solder-trap chamber. Continuous vacuum extraction is the only consistent method for overcoming the resweat problem for either dual or multilead devices terminating in through-hole solder joints.

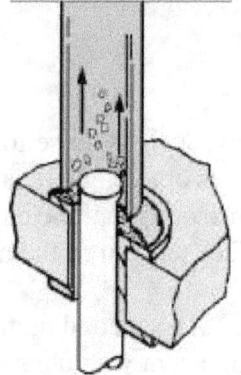

(A) VACUUM MODE
Figure 3-12A.—Motorized vacuum/pressure solder removal. VACUUM MODE.

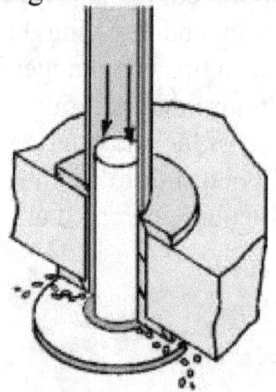

(B) PRESSURE MODE
Figure 3-12B.—Motorized vacuum/pressure solder removal. PRESSURE MODE.

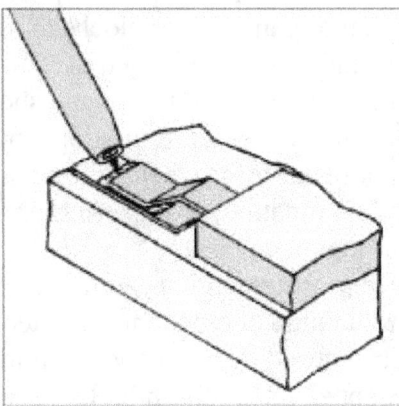

Figure 3-12C.—Motorized vacuum/pressure solder removal. HOT AIR JET MODE.

Motorized Vacuum Method.—In the motorized vacuum method, the heated tip is applied to the solder joint. When melted solder is observed, the vacuum is activated by the technician causing the solder to be withdrawn from the joint and deposited into the chamber. If the lead is preclipped, it may also be drawn into a holding chamber. To prevent SWEATING (reforming a solder joint) to the side walls of the plated-through hole joint, the lead is "stirred" with the tip while applying the vacuum. This permits cool air to flow into and around the lead and side walls causing them to cool.

Motorized Pressure Method.—In the pressure method, the tip is used to apply heat to a pin for melting a sweat joint. The air pressure is forced through the hole to melt sweat joints without contacting

the delicate pad. This method is seldom used because it is not effective in preventing sweating of the lead to the hole nor for cooling the workpiece.

Hot-Air Jet Method.—The hot-air jet method uses pressure-controlled, heated air to transfer heat to the solder joint without physical contact from a solder iron. This permits the reflow of delicate joints while minimizing mechanical damage.

When the solder is removed from the lead and pad area, the technician can observe the actual condition of the lead contact to the pad area and the amount of the remaining solder joint. From these observed conditions, the technician can then determine a method of removing the component and lead.

With straight-through terminations, the component and lead may be lifted gently from uncoated boards with pliers or tweezers. Working with clinched leads on uncoated boards requires that all sweat joints be removed and that the leads be unclinched before removal.

The techniques that have been described represent the successful methods of desoldering components. As mentioned at the beginning of this section, the 2M technician must decide which method is best suited for the type of solder joint. Two commonly used but unacceptable methods of solder removal are heat-and-shake and heat-and-pull methods.

In the heat-and-shake method, the solder joint is melted and then the molten solder is shaken from the connection. In some cases, the shaking action may include striking the assembly against a surface to shake the molten solder out of the joint. This method should NEVER be used because all the solder may not be removed and the solder may splatter over other areas of the board. In addition, striking the board against a surface can lead to broken boards, damaged components, and lifted pads or conductors.

The heat-and-pull method uses a soldering iron or gang-heater blocks to melt individual or multiple solder joints. The component leads are pulled when the solder is melted. This method has many shortcomings because of potential damage and should NOT be attempted. Heating blocks are patterned to suit specific configurations; but when used on multiple-lead connections, the joints may not be uniformly heated. Uneven heating results in plated-through hole damage, pad delamination, or blistering. Damage can also result when lead terminations are pulled through the board.

When desoldering is complete, the workpiece must undergo a careful physical inspection for damage to the circuit board and the remaining components. The technician should also check the board for scorching or charring caused by component failure. Sometimes MEASLING is present. Measling is the appearance of light-colored spots. It is caused by small areas of fiberglass strands that have been damaged by epoxy overcuring, heat, abrasion, or internal moisture. No cracks or breaks should be visible in the board material. None of the remaining components should be cracked, broken, or show signs of overheating. The solder joints should be of good quality and not covered by loose or splattered solder, which may cause shorts. The technician should examine the board for nicked, cracked, lifted, or delaminated conductors and lifted or delaminated pads.

Q10. When does most printed circuit board damage occur?

Q11. What procedure involves the use of finely braided copper wire to remove solder? Q12. What is the most effective method of solder removal?

Q13. When, if at all, should the heat-and-shake or the heat-and-pull methods of solder removal be used?

INSTALLATION AND SOLDERING OF PRINTED CIRCUIT COMPONENTS

The 2M technician should restore the electronic assembly at least to the original manufacturer's standards. Parts should always be remounted or reassembled in the same position and with termination methods used by the original manufacturer. This approach ensures a continuation of the original reliability of the system.

High reliability connections require thoroughly cleaned surfaces, proper component lead formation and termination, and appropriate placement of components on the board. The following paragraphs describe the procedures for properly installing components on a board including the soldering of these components.

Termination Area Preparation

The termination areas on the board and the component leads are thoroughly cleaned to remove oxide, old solder, and other contaminants. Old or excess solder is removed by one of the desoldering techniques explained earlier in this topic. A fine abrasive, such as an oil-free typewriter eraser, is used to remove oxides. This is not necessary if the area has just been desoldered. All areas to be soldered are cleaned with a solvent and then dried with a lint-free tissue to remove cleaning residue.

Component Lead Preparation

Component leads are formed before installation. Both machine- and hand-forming methods are used to form the leads. Improper lead formation causes many repairs to be unacceptable. Damage to the SEALS (point where lead enters the body of the component) occurs easily during the forming process and results in component failure. Consequently, lead-forming procedures have been established. To control the lead-forming operation and ensure conformity and quality of repairs, the technician should ensure the following:

1. The component is centered between the holes, and component leads are formed with proper bend-radii and body seal-to-bend distance.

2. The possibility of straining component body seals during lead forming is eliminated.

3. Stress relief loops are formed without straining component seals while at the same time providing the desired lead-to-lead distances.

4. Leads are measured and formed for both horizontal and vertical component mounting.

5. Transistor leads are formed to suit standard hole spacing. Lead-Forming Specifications.

Component leads are formed to provide proper lead spacing.

• The minimum distance between the seal (where the lead enters the body of the component) and the start of the lead bend must be no less than twice the diameter of the lead, as shown in figure 3-13.

BODY
SEAL
LEAD DIAMETER
2 X DIAMETER

Figure 3-13.—Minimum distance lead bend to component body.

Leads must be approximately 90 degrees from their major axis to ensure free movement in hole terminations, as shown in figure 3-14.

LEAD AXIS
I > j i
i j j i " > *

LEADS FIT FREELY IN BOTH HOLES EACH LEAD BEND IS SO DEGREES

Figure 3-14.—Ideal lead formation.

• In lead-forming, the lead must not be damaged by nicking.

• Energy from the bending action must not be transmitted into the component body.

COMPONENT PLACEMENT.—Where possible, parts are remounted or reassembled as they were in the original manufacturing process. To aid recognition, manufacturers use a coding system of colored dots, bands, letters, numbers, and signs. Replacement components are mounted to make all identification markings readable without disturbing the component. When components are mounted like the original, all the identification markings are readable from a single point.

Component identification reads uniformly from left to right, top to bottom, unless polarity requirements determine otherwise, as shown in figure 3-15. To locate the top, position the board so the part number may be read like a page in a book. By definition, the top of the board is the edge above the part number.

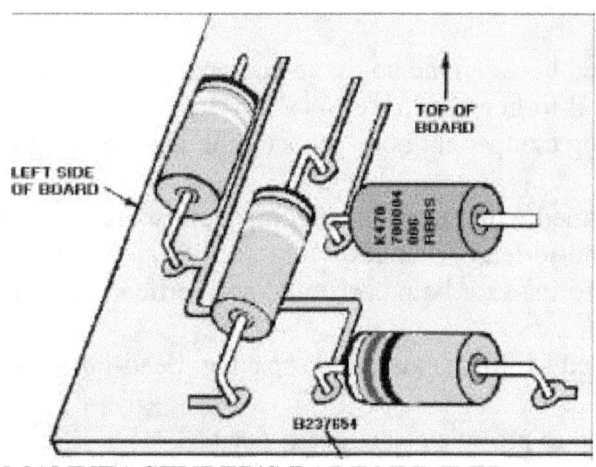

MANUFACTURER'S PART NUMBER

Figure 3-15.—Component arrangement.

When possible, component identification markings should be visible after installation. If you must choose between identification and electrical value markings, the priority of selection is as follows: (1) electrical value, (2) reliability level, and (3) part number.

Components are normally mounted parallel to and on the side opposite the printed circuitry and in contact with the board.

FORMATION OF PROPER LEAD TERMINATION —After component leads are formed and inserted into the board, the proper lead length and termination are made before the lead is soldered. Generally, if the original manufacturer clinched (either full or semi) the component leads, the replacement part is reinstalled with clinched leads.

When clinching is required, leads on single- and double-sided boards are securely clinched in the direction of the printed wiring connected to the pad. Clinching is performed with tools that prevent damage to the pad or printed wiring. The lead is clinched in the direction of the conductor by bending the lead. The leads are clipped so that their minimum clinched length is equal to the radius of the pad. Under no circumstances does the clinched lead extend beyond the pad diameter. Natural springback away from the pad or printed wiring is acceptable. A gap between the lead end and the pad or printed wiring is acceptable when further clinching endangers the pad or printed wiring. These guidelines ensure uniform lead length.

Q14. To what standards should a technician restore electronic assemblies?

Q15. How is oxide removed from pads and component leads?

Q16. Leads are formed approximately how many degrees from their major axis?

Ql 7. When you replace components, identification marks must meet what requirements?

Q18. In what direction are component leads clinched on single- and double-sided boards?

Soldering of PCB Components

The fundamental principles of solder application must be understood and observed to ensure consistent and satisfactory results. As discussed in topic 2, the soldering process involves a metal-solvent action that joins two metals by dissolving a small amount of the metals at their point of contact.

SOLDERABILITY.—As the solder interacts with the base metals, a good metallurgical bond is obtained and metallic continuity is established. This continuity is good for electrical and heat conductivity as well as for strength. Solderability measures the ease with which molten solder wets the surfaces of the metals being joined. WETTING means the molten solder leaves a continuous permanent film on the metal surface. Wetting can only be done properly on a clean surface. All dirt and grease must be removed and no oxide layer must exist on the metal surface. Using abrasives and/or flux to remove these contaminants produces highly solderable surfaces.

HEAT SOURCE.—The soldering process requires sufficient heat to produce alloy- or metal-solvent action. Heat sources include CONDUCTIVE, RESISTIVE, CONVECTIVE, and RADIANT types. The type of heat source most commonly used is the conductive-type soldering iron. Delicate electronic assemblies require that the thermal characteristics of a soldering iron be carefully balanced and that the iron and tip be properly matched to the job. Successful soldering depends on the combination of the iron tip temperature, the capacity of the iron to sustain temperature, the time of iron contact with the joint, and the relative mass and heat transfer characteristics of the object being soldered.

SELECTION OF PROPER TIP.—The amount of heat and how it is controlled are critical factors to the soldering process. The tip of the soldering iron transfers heat from the iron to the work. The shape and size of the tip are mainly determined by the type of work to be performed. The tip size and the wattage of the element must be capable of rapidly heating the mass to the melting temperature of solder.

After the proper tip is selected and attached to the iron, the operator may control the heat by using the variable-voltage control. The most efficient soldering temperature is approximately 550 degrees Fahrenheit. Ideally, the joint should be brought to this temperature rapidly and held there for a short period of time. In most cases the soldering action should be completed within 2 or 3 seconds. When soldering a small-mass connection, control the heat by decreasing the size of the tip.

Before heat is applied to solder the joint, a thermal shunt is attached to sensitive component leads (diodes, transistors, and ICs). A thermal shunt is used to conduct heat away from the component. Because of its large heat content and high thermal conductivity, copper is usually used to make thermal shunts. Aluminum also has good conductivity but a smaller heat content; it is also used to conduct heat, especially if damage from the physical weight of the clamp is possible. Many types, shapes, and sizes of thermal shunts are available. The most commonly used is the clamp design; this is a spring clip (similar to an alligator clip) that easily fastens onto the part lead, as shown in figure 3-16.

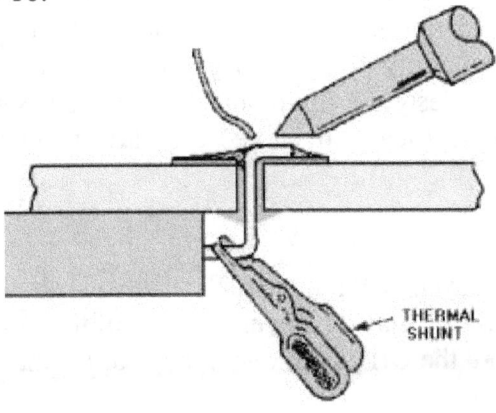

THERMAL
SHUNT

Figure 3-16.—Thermal shunt.

APPLICATION OF SOLDER AND SOLDERING IRON TIP —Before solder is applied to the joint, the surface temperature of the parts being soldered is increased above the solder melting point. In general, the soldering iron is applied to the point of greatest mass at the connection. This increases the heat in the parts to be soldered. Solder is then applied to a clean, fluxed, and properly heated surface. When properly applied, the solder melts and flows without direct contact with the heat source and provides a smooth, even surface that feathers to a thin edge.

Molten solder forms between the tip and the joint, creating a heat bridge or thermal linkage. This heat bridge causes the tip to become part of the joint and allows rapid heat transfer. A solder (heat) bridge is formed by melting a small amount of solder at the junction of the tip and the mass being soldered as the iron is applied. After the tip makes contact with the lead and the pad and after the heat bridge is established, the solder is applied with a wiping motion to form the solder bond. The completed solder joint should be bright and shiny in appearance. It should have no cracks or pits, and the solder should cover the pad. Examples of preferred solder joints are shown in figure 3-17. They are referred to as full fillet joints.

FULL FILLET
FULL FILLET
V
FULL FILLET

Figure 3-17.—Preferred solder joint.

When a solder joint is completed, solvent must be used to remove all flux residue. The two most highly recommended solvents, in the order of their effectiveness, are 99.5 percent pure ethyl alcohol and 99.5 percent pure isopropyl alcohol.

Q19. What is solderability?

Q20. What is the most common source of heat in electronic soldering?

Q21. What determines the shape and size of a soldering iron tip?

Q22. What term describes a device used to conduct heat away from a component?

Q23. What is the appearance of a properly soldered joint?

REMOVAL AND REPLACEMENT OF DIPS

In topic 1 you learned the advantages of DIPs. They are easily inserted by hand or machine and require no special spreaders, spacers, insulators, or lead-forming tools. Standard hand tools and soldering equipment can be used to remove and replace DIPS.

DIPs may be mounted on a board in two ways: (1) They may be mounted by plugging them into DIP mounting sockets that are soldered to the printed circuit boards or (2) they are soldered in place and may or may not be conformally coated. Although plug-ins are very easy to service, they lack the reliability of soldered-in units, do not meet MILSPECS, and are seldom used in military designed equipment. They are susceptible to loosening because of vibration and to poor electrical contact because of dust and dirt and corrosion.

3-25

Removal of Plug-In DIPs

To remove plug in DIPs, use an approved DIP puller, such as the one shown in figure 3-18. The puller shown is a plastic device that slips over the ends of the DIP and lifts the DIP evenly out of the socket. Before the DIP is removed, the board is marked or a sketch is made of the DIP reference mark location; then the reference mark for the replacement part will be in the proper position. The DIP is grasped with the puller and gently lifted straight out of the socket. Lifting one side or one end first results in bent leads. If the removed DIP is to be placed back in the circuit, particular care is taken in straightening bent leads to prevent breaking. To straighten bent leads, the technician grasps the wide portion of the lead with one pair of smooth-jaw needle nose pliers; with another pair, the technician then bends the lead into alignment with the other leads. Tools used for lead straightening should be cleaned with solvent to remove contaminants.

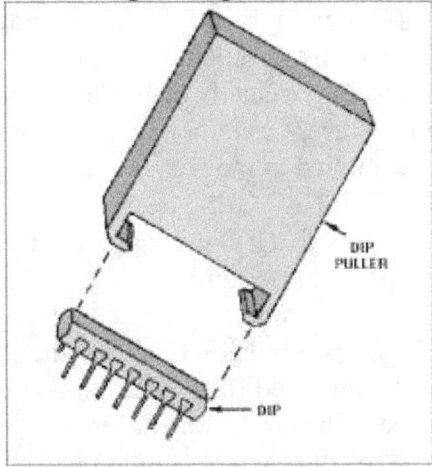

Figure 3-18.—Typical DIP puller.

To replace a plug-in DIP, the technician should clean the leads with solvent and then check the proper positioning of the reference mark. To do this, the technician holds the DIP body between the thumb and forefinger and places the part on the socket to check pin alignment. The pins are not touched. If all pins are properly aligned, the technician presses the part gently into the socket until the part is firmly seated. As pressure is applied, each pin is checked to ensure that all pins are going into the socket. If pins tend to bend, the part is removed and the pins are straightened. The socket is then inspected to make sure the holes are not obstructed. Then the process is repeated. After a thorough visual inspection, the card should be ready for testing.

Removal and Replacement of Soldered-In DIPs

The removal of soldered-in DIPs without conformal coatings is essentially the same as the removal of discrete components, except that a skipping pattern is always used. A skipping pattern is one that skips from pad to pad, never heating two pads next to each other. This reduces heat accumulation and reduces

the chance of damage to the board. Of course, many more leads should be desoldered before the part can be removed. Special care must be exercised to make sure all leads are completely free before an attempt is made to lift the part off the board. If the part is known to be faulty, or if normal removal may damage the board, then the leads should be clipped. Once this has been done, desoldering can be done from both sides of the board. After the clipped leads have been desoldered, they can be removed with tweezers or pliers.

The removal of DIPs from boards with conformal coatings should be completed in the same manner as for other components. The coating should be removed using the preferred method of removal for that particular type of material. The coating should be removed from both sides of the board after masking off the work area. Particular care should be taken when removing the material from around the delicate leads. If the part is to be reused, as much of the coating should be removed from the leads as possible. As with DIPs without conformal coatings, if the part is known to be bad or if the possibility of board damage exists, the leads are clipped; the part and leads are then removed as described earlier in this section. Once the part has been removed, the work area should be completely cleaned to remove any remaining coating or solder.

The steps for replacing a soldered-in DIP are similar to those for replacing a plug-in DIP. Once the part is in position, it is soldered using the same standard used by the manufacturer, or as close to that standard as is possible with the available equipment. The joints should be soldered as quickly as possible using only as much heat as is necessary using a skipping pattern. The repaired card should then be visually inspected for defects in workmanship, and testing of the card should take place. Once the successful repair has been accomplished, a conformal coating should be applied to the work area.

REMOVAL AND REPLACEMENT OF TO PACKAGES

You should recall from chapter 1 of this module, that TO packages are mounted in two ways— plugged-in or embedded. The term plug-in, when referring to TOs, should not be confused with DIP plug-ins. TOs are normally soldered in place. You will come across sockets for TOs, but not as frequently as for DIPs. Figure 3-19 shows the methods of mounting TOs. Notice that plug-ins may either be mounted flush with the board surface or above the surface with or without a spacer. The air gap or spacer may be used by the manufacturer for a particular purpose. This type of mounting could be used for heat dissipation, short circuit protection, or to limit parasitic interaction between components. The spacer also provides additional physical support for the TO. The technician is responsible for using the same procedure as the manufacturer to replace TOs or any other components.

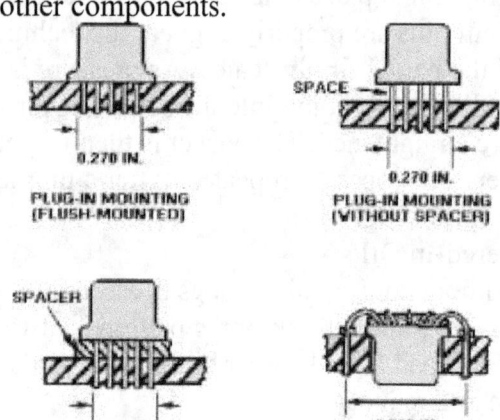

PLUG-IN MOUNTING EMBEDDED MOUNTING
(VITH SPACER)
Figure 3-19.—TO mounting techniques.

The procedure for removal of plug-in TOs (with or without conformal coatings) is the same as that used for a similarly mounted DIP or discrete component. The conformal coating is removed if required. Leads are desoldered and gently lifted out of the board. Then board terminals and component leads are cleaned.

In some plug-ins, the leads must be formed before they are placed in a circuit. Care should be taken to ensure that seal damage does not occur and that formed leads do not touch the TO case. This would result in a short-circuit.

When the new part or the one that was removed is installed, the leads are slipped through the spacer if required, and the part is properly positioned (reference tab in the proper location). The leads are aligned with the terminal holes and gently pressed into position. The part is soldered into place and visually inspected. Then the card is tested and the conformal coating is replaced if required.

The removal of an imbedded TO package varies only slightly from the removal of other types of mountings. First, the work area is masked and the conformal coating is removed if required. Then the desoldering handpiece is used to remove the solder from each lead. When all leads are free, the TO is pushed out of the board. If all the leads are free, the TO should slip out of the board easily. The package should not be forced out of the board. Excessive pressure may cause additional damage. If the leads are not completely free, the leads must be clipped and removed after the package is out of the board. This process is shown in figure 3-20.

© ©
PUSH OUT TO -5
CLIP LEADS
© ©
REMOVE LEAO ENOS
UNSOLDER LEADS
Figure 3-20.—Imbedded TO removal.

The most critical part of replacing an imbedded TO is the lead formation. The leads are formed to match the original part as closely as possible. Once the body and leads are seated, the leads can be soldered and the board inspected.

REMOVAL AND REPLACEMENT OF FLAT PACKS

Up to this point, all of the components discussed have had through-the-board leads. In addition, the removal and replacement of discrete components, DIPs, and TOs have been similar.

PLANAR-MOUNTED COMPONENTS (FLAT-PACKS)

Different techniques are used in the removal and replacement of flat packs and devices with on-the-board terminations. Lap-flow solder joints require that the technician pay particular attention to workmanship. Some of the standards of workmanship will be discussed later in this section.

Flat-Pack Removal

Prior to the removal of a flat pack, as with other ICs, a sketch should be prepared to identify the proper positioning of the part. The conformal coating should be removed as required.

To remove the flat pack, the 2M technician carefully heats the leads and lifts them free with tweezers. If the part is to be reused, special care is taken not to damage or bend the leads. The work area around the component should then be thoroughly cleaned and prepared for the new part.

Flat-Pack Replacement

Flat packs attached to boards normally have formed and trimmed leads. Manufacturers form and trim the leads in one operation with a combination die. However, most replacement flat packs are received in a protective holder (figure 3-21) and the leads must be formed and trimmed by hand. Cost prevents equipping the repair station with the variety of tools and dies to form leads because of the variety of component configurations.

Figure 3-21.—Flat pack in protective holder.

LEAD-BENDING TECHNIQUES.—The 2M technician learns several methods of lead forming that will provide proper contact for soldering and circuit operations. The techniques used to bend leads include the use of specialized tools and such common items as flat toothpicks, bobby pins, and excess component leads. Care is taken not to stress the seal of the component during any step of the lead forming. Figure 3-22 illustrates two views, view (A) and view (B), of properly formed flat-pack leads.

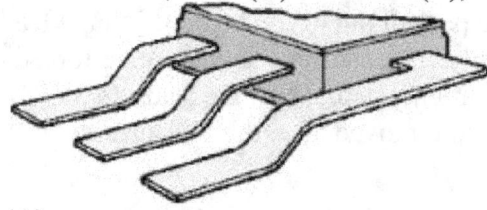

(A)

Figure 3-22A.—Properly formed flat pack leads.

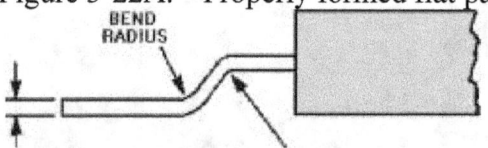

LEAD THICKNESS BEND
RADIUS

LEAD BEND RADII SHALL BE EQUAL TO OR GREATER THAN TWICE THE LEAD THICKNESS.

(B)

Figure 3-22B.—Properly formed flat pack leads.

Because most replacement flat packs come with leads that are longer than required, they must be trimmed before they are soldered. The removed part is used as a guide in determining lead length. Surgical scissors or scalpels are recommended for use in cutting flat-pack leads. Surgical scissors permit all leads to be cut to the required lead length in a smooth operation with no physical shock transmitted to the IC.

LAP-SOLDERING CONNECTIONS.—Before a connection is lap-soldered, the solder pads are cleaned and pretinned and the component leads are tinned. This is particularly important if they are gold plated. The IC is properly positioned on the pad areas, and the soldering process is a matter of "sweating" the two conductors together. When multilead components, such as ICs, are soldered, a skipping pattern is used to prevent excessive heat buildup in a single area of the board or component. When soldering is completed, all solder connections are thoroughly cleaned. All joints should be inspected and tested. The standards of workmanship are more specific for flat-pack installation.

Q24. When removing the component, under what circumstances may component leads be clipped?

Q25. How are imbedded TOs removed once the leads are free?

Q26. How is a flat pack removed from a pcb?

Q27. How do you prevent excessive heat buildup on an area of a board when soldering multilead components?

Q28. What are the two final steps of any repair?

REPAIR OF PRINTED CIRCUIT BOARDS AND CARDS

Removal and replacement of components on boards and circuit cards are, by far, the most common types of repair. Equally important is the repair of damaged or broken cards. Proper repair of damaged boards not only maintains reliability of the board but also maintains reliability of the system.

Cards and boards may be damaged in any of several ways and by a number of causes. Untrained personnel making improper repairs and technicians using improper tools are two major causes of damage. Improper shipping, packaging, storage, and use are also common sources of damage. The source of damage most familiar to technicians is operational failure. Operational failures include cracking caused by heat, warping, component overheating, and faulty wiring.

Before attempting board repairs, the technician should thoroughly inspect the damage. The decision to repair or discard the piece depends on the extent of damage, the level of maintenance authorized, operational requirements, and the availability of repair parts and materials. The following procedures will help you become familiar with the steps necessary to repair particular types of damage. Remember, only qualified personnel are authorized to attempt these repairs.

Repair of Conductor and Termination Pads

Conductor (run) and pad damage is very common. The technician must examine the board for nicks, tears, or scratches that have not broken the circuit, as well as for complete breaks, as shown in figure 3-23. Crack damage may exist as nicks or scratches in the conductor. These nicks or scratches must be repaired if over one-tenth of the cross-sectional area of the conductor is affected as current-carrying capability is reduced. Cracks may also penetrate the conductor.

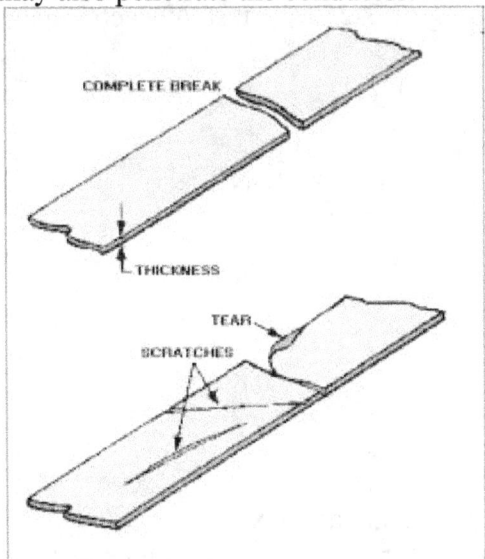

Figure 3-23.—Pcb conductor damage.

CRACK REPAIR.—Four techniques are used to repair cracks in printed circuit conductors. One method is to flow solder across the crack to form a solder bridge. This is not a high-reliability repair since the solder in the break will crack easily.

The second method is to lap-solder a piece of wire across the crack. This method produces a stronger bond than a solder bridge; but it is not highly reliable, as the solder may crack.

A third repair technique is to drill a hole through the board where the crack is located and then to install an eyelet in the hole and solder it into place.

The fourth method is to use the clinched-staple method, shown in figure 3-24. It is the most reliable method and is recommended in nearly all cases.

DAMAGED CONDUCTOR (BREAK)
CLINCH LEAD AGAINST CONDUCTOR AND SOLDER
1M6"min. TYPICAL
\
W8" MAX. INCLUDING SOLDER
3^

LAMINATED BOARD
EQUIV. GAUGE SOLID TINNED WIRE WITH SLEEVING

Figure 3-24.—Clinched-staple repair of broken conductor.

Pads or conductor runs may be completely missing from the board. These missing pads or runs must be replaced. Also included in this type of damage are conductors that are present but damaged beyond repair.

REPLACING DAMAGED OR MISSING CONDUCTORS —The procedures used to replace damaged or missing conductors are essentially the same as using the clinched-staple method of conductor repair.

REPLACING THE TERMINATION PAD.—Many times the termination pad, as well as part of the conductor, is missing on the board. In these cases, a replacement pad is obtained from a scrap circuit board. Refer to figure 3-25 as you study each step.

VIEV OF UNDERSIDE

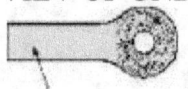

AREA CLEANED OF EPOXY & CONTAMINATION
i A MINIMUM OF 1 2X VODTH

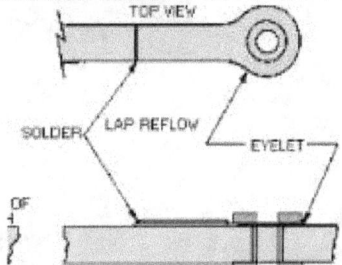

Figure 3-25.—Replacement of damaged termination pad.

The underside of the replacement pad and the area where it will be installed is cleaned. An epoxy is used to fasten the replacement pad to the board. An eyelet is installed to reinforce the pad before the epoxy sets and cures. This ensures a good mechanical bond between the board and pad and provides good electrical contact for components. After the epoxy cures, the new pad is lap-soldered to the original run.

REPAIRING DELAMINATED CONDUCTORS.—DELAMINATED CONDUCTORS (figure 3-26) are classified as conductors no longer bonded to the board surface. Separation of the laminations may occur only on a part of the conductor. Proper epoxying techniques ensure complete bonding of the conductor to the circuit board laminate. The following procedures are used to obtain a proper bond:

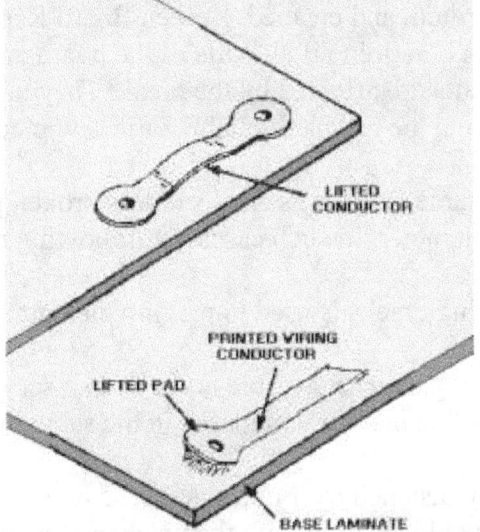

Figure 3-26.—Delaminated conductors.

1. A small amount of epoxy is mixed and applied to the conductor and the conductor path; no areas are left uncoated.

2. The conductor is clamped firmly against the board surface until the epoxy has completely cured.

REPLACING EYELETS.—Eyelets have been referred to in several places in this topic. Not only are they used for through-the-board terminations, but also to reinforce some types of board repairs. As with any kind of material, eyelets are subject to damage. Eyelets may break, they may be installed improperly, or they may be missing from the equipment. When an eyelet is missing or damaged, regardless of the kind of damage, it should be replaced. The guidelines for the selection and installation of new eyelets are far too complex to explain here. However, they do comprise a large part of the 2M technician's training.

Repair of Cracked Boards

When boards are cracked, the length and depth of the cracks must be determined. Also, the disruption to conductors and components caused by cracks must be determined by visual inspection. To avoid causing additional damage, the technician must exercise care when examining cracked boards and

must not flex the board. Rebuilding techniques must be used to repair damage, such as cracks, breaks, and holes that extend through the board. The following steps are used to repair cracks:

1. Abrasive methods are used to remove all chips and fractured material.

2. The edges of the removed area are beveled and undercut to provide bond strength.

3. A smoothly surfaced, nonporous object is fastened tightly against one side of the removed area.

4. The cutaway area is filled with a compound of epoxy and powdered fiberglass (figure 3-27). Extreme care is exercised to prevent the formation of voids or air bubbles in the mixture.

Figure 3-27.—Repair of cracked pcbs.

5. The surface of the filled area is smoothed to make it level with the surface of the original board.

6. The board is cured, smoothed, redrilled, and cleaned. Broken Board Repair

Broken boards should be examined to determine if all parts of the board are present and if circuit conductors or components are affected by the break. They are also examined to determine if the broken pieces may be rejoined reliably or if new pieces must be manufactured.

Breaks and holes are repaired in the same manner as cracks unless broken pieces are missing or the hole exceeds 1/2 inch in diameter. In such cases, the following repair steps are used:

1. The same technique used in repairing cracks is used to prepare the damaged edge.

2. A piece as close in size to the missing area as possible is cut from a scrap board of the same type and thickness. The edges of this piece are prepared in the same manner as the edges of the hole.

3. A smooth-surfaced object is tightly fastened over one side of the repair area, and the board is firmly clamped in an immovable position with the uncovered area facing up.

4. The replacement piece is positioned as nearly as possible to the original board configuration and firmly clamped into place.

5. The repair is completed using the same epoxy-fiberglass mixture and repair techniques used in the patching repair method discussed in the following section on burned board repair.

Burned Board Repair

Scorched, charred, or deeply burned boards should be inspected to determine the size of the discolored area and to identify melted or blackened conductors and burned, melted, or blackened components. The depth of the damage, which may range from a slight surface discoloration to a hole burned through the circuit board, should also be determined. Damage not extending through the board may be repaired by patching (figure 3-28). The following procedure is used in the repair of these boards.

1. If the board is scorched, charred, or burned, all discolored board material is removed by abrasive methods, as shown in figure 3-29. Several components in the affected area may have to be desoldered and removed before the repair is continued.

Figure 3-28.—Repair of surface damage.
BURNED RESISTOR

DISCOLORED AREA i
CUTAWAY VIEV. DAMAGED BOARD

ALL CHARREDf DISCOLORED AREAS GROUND AWAY. EDGES UNDERCUT IO PROf IDE PHTSICAL HOLDING POINTS FOR THE REPAIR HART ERIAL

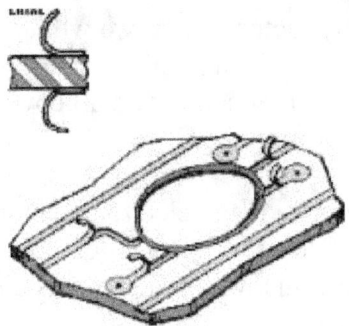

Figure 3-29.—Repair of burned boards.

2. Repairable delaminations not extending to the edge of the circuit board should be cut away by abrasive methods until no delaminated material remains.

3. Delaminated material is not removed if it is repairable.

4. After all damaged board material is removed, the edge of the removed area is beveled and undercut to provide holding points for the repair material.

5. Solvent is used to clean thoroughly and to remove all loose particles.

6. A compound of epoxy and powdered fiberglass is mixed and used to fill the cutaway area.

7. The epoxy repair mixture is cured according to the manufacturer's instructions.

8. The surface of the filled area is leveled after the compound is cured.

9. If delaminations extend to the edge of the board, the delaminated layers are filled completely with the repair mixture and clamped firmly together between two flat surfaces.

10. After the cure is completed, abrasive methods are used to smooth the repaired surface to the same level as the original board.

11. If necessary, needed holes are redrilled in the damaged area, runs are replaced, eyelets and components are installed, and the area is cleaned. Figure 3-30 shows the repaired area ready for components.

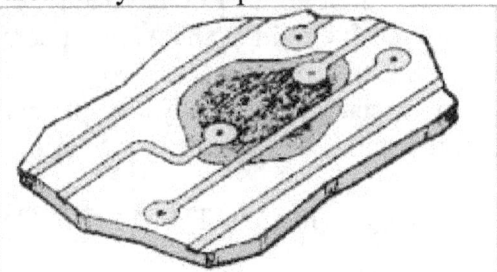

Figure 3-30.—Repaired board ready for components.

Q29. List three causes of damage to printed circuit boards.

Q30. What is the preferred method of repairing cracked runs on boards?

Q31. Damaged or missing termination pads are replaced using what procedure?

Q32. How is board damage caused by technicians?

Q33. What combination of materials is used to patch or build up damaged areas of boards?

SAFETY

Safety is a subject of utmost importance to all technical personnel. Potentially hazardous situations exist in almost any work area. The disregard of safety precautions can result in personal injury or in the loss of equipment or equipment capabilities.

In this section we will discuss two types of safety factors. First, we will cover damage that can occur to electronic components because of electrostatic discharge (ESD) and improper handling and stowage of parts and equipment. Second, we will cover personal safety precautions that specifically concern the technician.

ELECTROSTATIC DISCHARGE

Electrostatic discharge (ESD) can destroy or damage many electronic components including integrated circuits and discrete semiconductor devices. Certain devices are more susceptible to ESD damage than others. Because of this, warning symbols are now used to identify ESD-sensitive (ESDS) items (figure 3-31).

Figure 3-31.—Warning symbols for ESDS devices.

Static electricity is created whenever two substances (solid or fluid) are rubbed together or separated. This rubbing or separation causes the transfer of electrons from one substance to the other; one substance then becomes positively charged and the other becomes negatively charged. When either of these charged substances comes in contact with a conductor, an electrical current flows until that substance is at the same electrical potential as ground.

You commonly experience static build-up during the winter months when you walk across a vinyl or carpeted floor. (Synthetics, especially plastics, are excellent generators of static electricity.) If you then touch a door knob or other conductor, an electrical arc to ground may result and you may receive a slight shock. For a person to experience such a shock, the electrostatic potential created must be 3,500 to 4,000 volts. Lesser voltages, although present and similarly discharged, normally are not apparent to a person's nervous system. Some typical measured static charges caused by various actions are shown in table 3-2.

Table 3-2.—Typical Measured Statics Charges (in volts) ITEM RELATIVE HUMIDITY

ITEM	LOW (10-20%)	HIGH (65-90%)
WALKING ACROSS CARPET	35,000	1,500
WALKING OVER VINYL FLOOR	12,000	250
WORKER AT BENCH	6,000	100
VINYL ENVELOPES FOR WORK INSTRUCT.	7,000	600
POLY BAG PICKED UP FROM BENCH	20,000	1,200
WORK CHAIR PADDED WITH URETHANE FOAM	18,000	1,500

Metal oxide semiconductor (MOS) devices are the most susceptible to damage from ESD. For example, an MOS field-effect transistor (MOSFET) can be damaged by a static voltage potential of as little as 35 volts. Commonly used discrete bipolar transistors and diodes (often used in ESD-protective circuits), although less susceptible to ESD, can be damaged by voltage potentials of less than 3,000 electrostatic volts. Damage does not always result in sudden device failure but sometimes results in device degradation and early failure. Table 3-2 clearly shows that electrostatic voltages well in excess of 3,000 volts can be easily generated, especially under low-humidity conditions. ESD damage of ESDS parts or circuit assemblies is possible wherever two or more pins of any of these devices are electrically exposed or have low impedance paths. Similarly, an ESDS device in a printed circuit board, or even in another pcb that is electrically connected in a series can be damaged if it provides a path to ground. Electrostatic discharge damage can occur during the manufacture of equipment or during the servicing of the equipment. Damage can occur anytime devices or assemblies are handled, replaced, tested, or inserted into a connector.

Technicians should be aware of the many sources of static charge. Table 3-3 lists many common sources of electrostatic charge. Although they are of little consequence during most daily activity, they become extremely important when you work with ESD material.

Table 3-3.—Common Sources of Electrostatic Charge

Prevention of ESD Damage

Certified 2M technicians are trained in procedures for reducing the causes of ESD damage. The procedures are similar for all levels of maintenance. The following procedure is an example of some of the protective measures used to prevent ESD damage.

1. Before starting to service equipment, the technician should be grounded to discharge any static electric charge built up on the body. This can be accomplished with the use of a test lead (a single-wire conductor with a series resistance of 1 megohm equipped with alligator clips on each end). One clip end is connected to the grounded equipment frame, and the other clip end is

touched with a bare hand. Figure 3-32 shows a more refined ground strap which frees both hands for work.

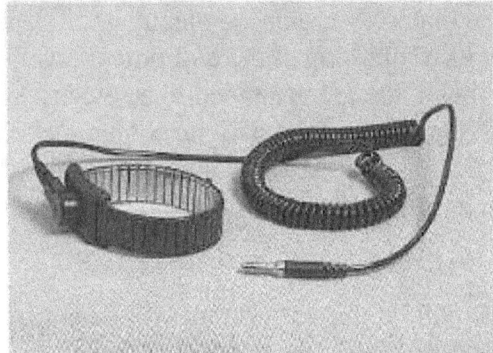

Figure 3-32.—ESD wrist strap.

2. Equipment technical manuals and packaging material should be checked for ESD warnings and instructions.

3. Prior to opening an electrostatic unit package of an electrostatic sensitive device or assembly, clip the free end of the test lead to the package. This will cause any static electricity which may have built up on the package to discharge. The other end remains connected to the equipment frame or other ESD ground. Keep the unit package grounded until the replacement device or assembly is placed in the unit package.

4. Minimize handling of ESDS devices and assemblies. Keep replacement devices or assemblies, with their connector shorting bars, clips, and so forth, intact in their

electrostatic-free packages until needed. Place removed repairable ESD devices or assemblies with their connector shorting bars/clips installed in electrostatic-free packages as soon as they are removed from the equipment. ESDS devices or assemblies are to be transported and stored only in protective packaging.

5. Always avoid unnecessary physical movement, such as scuffing the feet, when handling ESDS devices or assemblies. Such movement will generate additional charges of static electricity.

6. When removing or replacing an ESDS device or assembly in the equipment, hold the device or assembly through the electrostatic-free wrap if possible. Otherwise pick up the device or assembly by its body only. Do not touch component leads, connector pins, or any other electrical connections or paths on boards, even though they are covered by conformal coating.

7. Do not permit ESDS devices or assemblies to come in contact with clothing or other ungrounded materials that could have an electrostatic charge. The charges on a nonconducting material are not equal. A plastic storage bag may have a -10,000 volt potential 1/2 inch from a +15,000 volt potential, with many such charges all over the bag. Placing a circuit card inside the bag allows the charges to equalize through the pcb conductive paths and components, thereby causing failures. Do not hand an ESD device or assembly to another person until the device or assembly is protectively packaged.

8. When moving an ESDS device or assembly, always touch (with bare skin) the surface on which it rests for at least one second before picking it up. Before placing it on any surface, touch the surface with your free hand for at least one second. The bare skin contact provides a safe discharge path for charges accumulated while you are moving around.

9. While servicing equipment containing ESD devices, do not handle or touch materials such as plastic, vinyl, synthetic textiles, polished wood, fiberglass, or similar items which create static charges; or, be sure to repeat the grounding action with the bare hands after contacting these materials. These materials are prime electrostatic generators.

10. If possible, avoid repairs that require soldering at the equipment level. Soldering irons must have heater/tips assemblies that are grounded to ac electrical ground. Do not use ordinary plastic solder suckers (special antistatic solder suckers are commercially available).

11. Ground the leads of test equipment momentarily before you energize the test equipment and before you probe ESD items.

Grounded Work Benches

Work benches on which ESDS items will be placed and that will be contacted by personnel should have ESD protective work surfaces. These protective surfaces should cover the areas where ESD items will be placed. Personnel ground straps are also necessary for ESD protective work bench surfaces. These straps prevent people from discharging a static charge through an ESDS item to the work bench surface. The work bench surface should be connected to ground through a ground cable. The resistance in the bench top ground cable should be located at or near the point of contact with the work bench top. The resistance should be high enough to limit any leakage current to 5 milliamperes or less; this is taking into consideration the highest voltage source within reach of grounded people and all parallel resistances to ground, such as wrist ground straps, table tops, and conductive floors. See figure 3-33 for a typical ESD ground work bench.

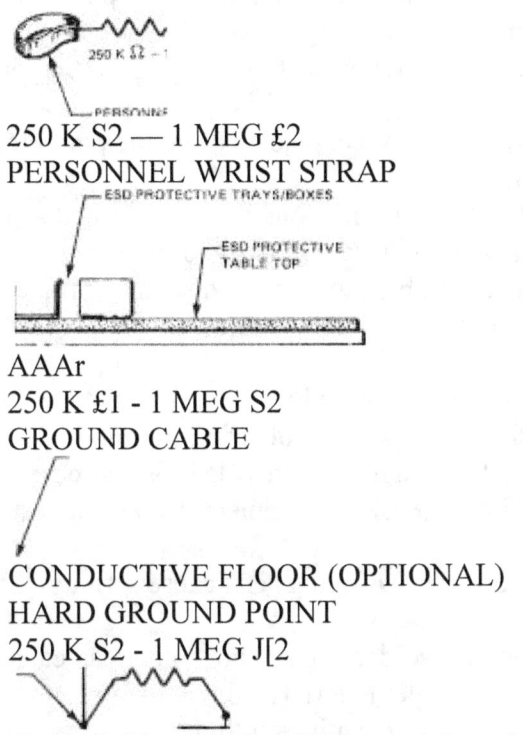

250 K S2 — 1 MEG £2
PERSONNEL WRIST STRAP

AAAr
250 K £1 - 1 MEG S2
GROUND CABLE

CONDUCTIVE FLOOR (OPTIONAL)
HARD GROUND POINT
250 K S2 - 1 MEG J[2

Figure 3-33.—Typical ESD ground work bench.

Energized equipment provides protection from ESD damage through operating circuitry. Circuit cards with ESD sensitive devices are generally considered safe when installed in an equipment rack; but they may be susceptible to damage if a "drawer" or "module" is removed and if connector pins are touched (even putting on plastic covers can transfer charges that do damage). There must not be any energized equipment placed on the conductive ESD work surface. An ESD work area is for "dead" equipment ONLY.

ESD protection is critical. If you should be assigned to 2M repair school, your education in ESD prevention will be quite extensive.

PERSONAL SAFETY

Throughout your career you will be aware of emphasis placed on safety. Safety rules remind you of potential dangers in work. Most accidents are preventable. Accidents don't happen without a cause. Most accidents are the result of not following prescribed safe operating procedures.

This would be a good time to review the safety section in topic 5 of NEETS, Module 2, Introduction to Alternating Current and Transformers. That section covers the basics of electrical shock and how to prevent it.

The 2M technician should be aware of other potential dangers in addition to the dangers of electrical shock. These dangers are discussed in the following paragraphs.

Power Tools

Hazards associated with the use of power tools include electrical shock, cuts, and particles in the eye. Safe tool use practices reduce or eliminate such accidents. Listed below are some of the general safety precautions that you should observe when your work requires the use of power tools.

• Ensure that all metal-cased power tools are properly grounded.

• Do not use spliced cables unless an emergency warrants the risks involved.

• Inspect the cord and plug for proper connection. Do not use any power tool that has a frayed cord or broken or damaged plug.

• Make sure that the on/off switch is in the OFF position before inserting or removing the plug from the receptacle.

• Always unplug the extension cord from the receptacle before the portable power tool is unplugged from the extension cord.

• Ensure all cables are positioned so they will not constitute a tripping hazard.

• Wear eye protection (goggles) in work areas where particles may strike the eye.

• After completing a task requiring a portable power tool, disconnect the power cord as described above and store the tool in its assigned location.

Soldering Iron

When using a soldering iron, remember the following:

• To avoid burns, always assume that a plugged-in soldering iron is HOT.

• Never rest a heated iron anywhere but in a holder provided for that purpose. Faulty action on your part could result in fire, extensive equipment damage, and/or serious injuries.

• Never use an excessive amount of solder. Drippings can cause serious skin or eye burns and can cause short circuits.

• Do not swing an iron to remove excess solder. Bits of hot solder can cause serious skin or eye burns or may ignite combustible material in the work area.

• When cleaning an iron, use a natural fiber cleaning cloth; never use synthetics, which melt. Do not hold the cleaning cloth in your hand. Always place the cloth on a suitable surface; then wipe the iron across it to avoid burning your hand.

• Hold small soldering jobs with pliers or a suitable clamping device to avoid burns. Never hold the work in your hand.

• Do not use an iron that has a frayed cord or damaged plug.

• Do not solder electronic equipment unless the equipment is electrically disconnected from the power supply circuit.

• After completing a task requiring a soldering iron other than the iron that is part of a work station, disconnect the power cord from the receptacle. When the iron has cooled, store it in its assigned stowage area.

Cleaning Solvents

The technician who smokes while using a cleaning solvent is inviting disaster. Unfortunately, many such disasters have occurred. For this reason, the Navy does not permit the use of gasoline, benzine, ether, or like solvents for cleaning since they present potential fire or explosion hazards. Only nonvolatile solvents should be used to clean electrical or electronic apparatus.

In addition to the potential hazard of accidental fire or explosion, most cleaning solvents can damage the human respiratory system where the fumes are breathed for a period of time.

The following positive safety precautions should be followed when performing cleaning operations.

• Use a blower or canvas wind chute to blow air into a compartment in which a cleaning solvent is being used.

• Open all usable port holes and place wind scoops in them.

• Place a fire extinguisher nearby.

• If it can be done, use water compounds instead of other solvents.

• Wear rubber gloves to prevent direct contact with solvents.

• Use goggles when a solvent is being sprayed on surfaces.

• Hold the nozzle close to the object being sprayed.

Where water compounds cannot be used, inhibited methyl chloroform (1.1.1 trichloroethane) should be used. Carbon tetrachloride is not used. Cleaning solvents that end with ETHYLENE are NOT safe to use. Methyl chloroform is an effective cleaner and is as safe as can be expected when reasonable care is exercised, such as adequate ventilation and the observance of fire precautions. When using inhibited methyl chloroform, avoid direct inhalation of the vapor. It is not safe for use, even with a gas mask, because its vapor displaces oxygen in the air.

Aerosol Dispensers

A 2M technician will encounter several uses for aerosol dispensers. The most common type is in applying conformal coatings.

Specific instructions concerning the precautions and procedures that must be observed to prevent physical injury cannot be given in this section because of the many available industrial sprays. However, all personnel concerned with handling aerosol dispensers containing volatile substances must clearly understand the hazards involved. They must also understand the importance of exercising protective measures to prevent personal injury. Strict compliance with the instructions printed on the aerosol dispensers will prevent many accidents that result from misapplication, mishandling, or improper storage of industrial sprays.

The rules for safe use of aerosol dispensers are listed below:

• Carefully read and comply with the instructions printed on the container.

• Do not use any dispenser that is capable of producing dangerous gases or other toxic effects in an enclosed area unless the area is adequately ventilated.

• If a protective coating must be sprayed in an inadequately ventilated space, either an air respirator or a self-contained breathing apparatus should be provided. However, fresh air supplied from outside the enclosure by exhaust fans or portable blowers is preferred. Such equipment prevents inhalation of toxic vapors.

• Do not spray protective coating on warm or energized equipment because this creates a fire hazard.

• Avoid skin contact with the liquid. Contact with some liquids may cause burns, while milder exposure may cause rashes. Some toxic materials are actually absorbed through the skin.

• Do not puncture the dispenser. Because it is pressurized, injury can result.

• Keep dispensers away from direct sunlight, heaters, and other heat sources.

• Do not store dispensers in an environment where the temperature exceeds the limits printed on the can. High temperatures may cause the container to burst.

Q34. List two causes of damage to ESD-sensitive electronic components.

Q35. What is the purpose of the wrist ground strap?

Q36. What is the cause of most accidents?

SUMMARY

This topic has presented information on miniature and microminiature (2M) repair procedures and 2M safety precautions. The information that follows summarizes the important points of this topic.

CONFORMAL COATINGS are protective materials applied to electronic assemblies to prevent damage caused by corrosion, moisture, and stress.

CONFORMAL COATINGS REMOVAL is accomplished mechanically, chemically, or thermally, depending on the material used.

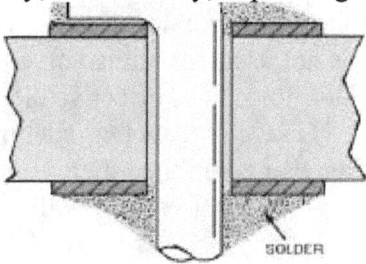

(A) FULLY CLINCHED

Component **LEADS** are terminated either through the board, above the board, or on the board.

SOLDER may be removed by wicking, by a manual vacuum plunger, or by a continuous vacuum solder extractor.

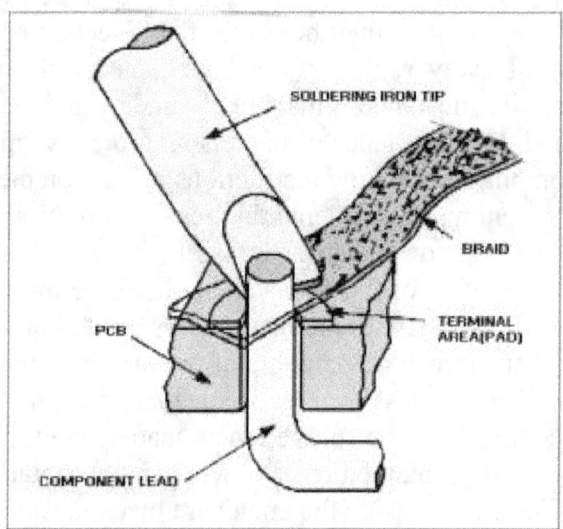

ELECTRONIC ASSEMBLIES should be restored to the original manufacturer's standards using the same orientation and termination method.

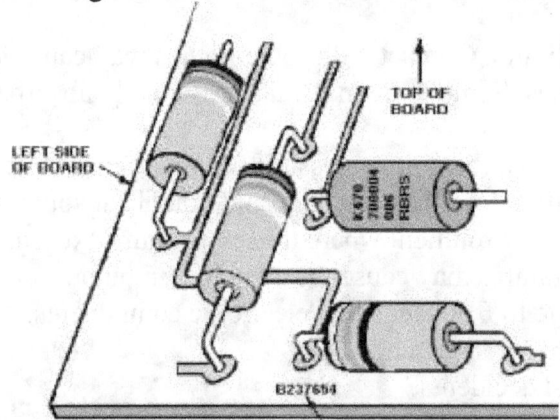

MANUFACTURER'S PART NUMBER

A GOOD SOLDER JOINT is bright and shiny with no cracks or pits.

When REPLACING DIPs, TOs, AND FLAT PACKS, make certain that pins are placed in the proper position.

COMPONENT LEADS may be clipped prior to removal only if the part is known to be bad or if normal removal will result in board damage.

The technician must determine through INSPECTION what method of repair is necessary for the board.

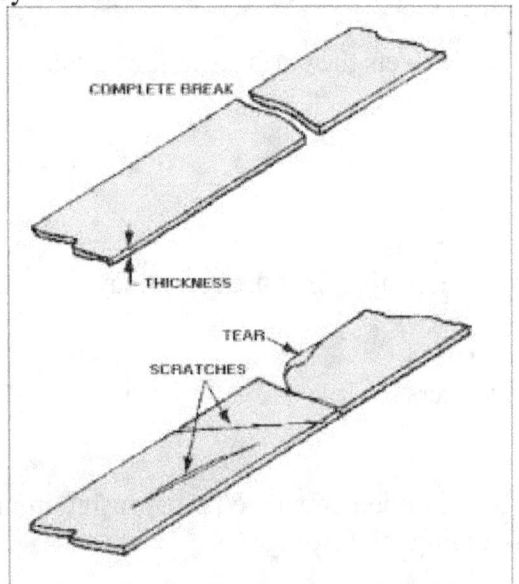

ELECTROSTATIC DISCHARGE (ESD) can damage or destroy many types of electronic components including integrated circuits and discrete components.

Special handling is required for ELECTROSTATIC-DISCHARGE-SENSITIVE (ESDS) devices or components.

USE PRESCRIBED SAFETY PRECAUTIONS when you use power tools, soldering irons, cleaning solvents, and aerosol dispensers.

ANSWERS TO QUESTIONS Ql. THROUGH Q36.

Al. Conformal coating.

A2. Chemical, mechanical, and thermal.

A3. Solvents or xylene and trichloroethane.

A4. Mechanical.

A5. To ensure protective characteristics are maintained.

A6. Interfacial connections.

A 7. Clinched lead, straight-through, and offset pad.

A8. Above-the-board termination.

A9. On-the-board termination.

A10. During disassembly or repair.

All. Wicking.

All. Continuous vacuum.

A13. These methods should not be used.

A14. Manufacturer's standards.

A15. A fine abrasive.

A16. 90 degrees.

A17. They should be readable from a single point.

A18. In the direction of the run.

A19. The ease with which molten solder wets the surfaces of the metals to be joined.

A20. Conductive-type soldering iron.

A21. The type of work to be done.

A22. A thermal shunt.

A23. Bright and shiny with no cracks or pits.

A24. If the component is known to be defective or if the board may be damaged by normal desoldering.

A25. By pushing it gently out of the board.

A26. Heat each lead and lift with tweezers.

A27. Use a skipping pattern.

A28. Inspect and test.

A29. Operational failures, repairs by untrained personnel, repair using improper tools, mishandling, improper shipping, packaging, and storage.

A30. Clinched staple.

A31. Epoxy a replacement pad to the board, set an eyelet, and solder it.

A32. Repairs by untrained personnel and technicians using improper tools.

A3 3. Epoxy and fiberglass powder.

A34. Esd, improper stowage, and improper handling.

A35. To discharge any static charge built up in the body.

A36. Deviation from prescribed safe operating procedures.

APPENDIX I

GLOSSARY

ALLOWANCE PARTS LIST (APL)—Repair parts required for unit having the equipment/ component listed.

ALLOWANCE EQUIPAGE LIST (AEL)—Equipment requirements for a unit having the exact equipment/component listed.

BEAM-LEAD CHIP—Semiconductor chip with electrodes (leads) extended beyond the wafer.

BONDING WIRES—Fine wires connecting the bonding pads of the chip to the external leads of the package.

BUILT-IN TEST EQUIPMENT (BITE)—Permanently mounted to the equipment for the purpose of testing the equipment.

CABLE HARNESS—A group of wires or ribbons of wiring used to interconnect electronic systems and subsystems.

CATHODE SPUTTERING—Process of producing thin film components.

CERMET—A combination of powdered precious-metal alloys and an inorganic material such as alumina. Used in manufacturing resistors, capacitors, and other components for high-temperature applications.

CORDWOOD MODULE.—A method of increasing the number of discrete components in a given space. Resembles wood stacked for a fireplace.

CRYSTAL FURNACE.—Device for artificially growing cylindrical crystals for producing semiconductor substrates.

DEPOT-LEVEL MAINTENANCE (SM&R Code D)—Supports S&R Code I and SM&R Code O activities through extensive shop facilities and equipment and more highly skilled personnel.

DICE—Uncased chips.

DIE BONDING—Process of mounting a chip to a package.

DIFFUSION—Controlled application of impurity atoms to a semiconductor substrate. DISCRETE COMPONENTS—Individual transistors, diodes, resistors, capacitors, and inductors. DOPING—See Diffusion.

DUAL IN-LINE PACKAGE (DIP)—IC package having two parallel rows of preformed leads. ENCAPSULATED—Imbedded in solid material or enclosed in glass or metal.

AI-1

EPITAXIAL PROCESS—The depositing of a thin uniformly doped crystalline region (layer) on a substrate.

EUTECTIC ALLOY—An alloy that changes directly from a solid to a liquid with no plastic or semiliquid state.

EUTECTIC SOLDER—An alloy of 63 percent tin and 37 percent lead. Melts at 361° F.

FILM ICs—Conductive or nonconductive material deposited on a glass or ceramic substrate. Used for passive circuit components, resistors, and capacitors.

FLAT PACK—IC package.

FLIP CHIP—Monolithic IC packaging technique that eliminates need for bonding wires. FLUX—Removes surface oxides from metals being soldered.

GENERAL PURPOSE ELECTRONIC TEST EQUIPMENT (GPETE)—Multimeters, oscilloscopes, voltmeters, signal generators, etc.

GROUND PLANES—Copper planes-used to minimize interference between circuits and from external sources.

HYBRID ICs—Two or more integrated circuit types, or one or more integrated circuit types and discrete components on a single substrate.

INTEGRATED CIRCUIT (IC)—Elements inseparably associated and formed on or within a single substrate.

INTERMEDIATE-LEVEL MAINTENANCE (SR&R Code I)—Direct support and technical assistance to user organizations. Tenders and shore-based repair facilities.

ISOLATION—The prevention of unwanted interaction or leakage between components.

LANDS—Conductors or runs on pcbs.

LARGE SCALE INTEGRATION (Isi)—An integrated circuit containing 1,000 to 2,000 logic gates or up to 64,000 bits of memory.

MASK—A device used to deposit materials on a substrate in the desired pattern.

MICROCIRCUIT—A small circuit having high equivalent-circuit-element density, which is considered as a single part composed of interconnected elements on or within a single substrate to perform an electronic-circuit function.

MICROELECTRONICS—That area of electronics technology associated with electronic systems built of extremely small electronic parts or elements.

MICROCIRCUIT MODULE—An assembly of microcircuits or a combination of microcircuits and discrete components that perform one or more distinct functions.

MODIFIED TRANSISTOR OUTLINE (TO)—IC package resembling a transistor.

MODULAR PACKAGING—Circuit assemblies or subassemblies packaged to be easily removed for maintenance or repair.

MODULE—A circuit or portion of a circuit packaged as a removable unit. A separable unit in a packaging scheme displaying regularity of dimensions.

MILITARY STANDARDS (MILSTD)—Standards of performance for components or equipment that must be met to be acceptable for military systems.

MINIATURE ELECTRONICS—Modules, packages, pcbs, and so forth, composed exclusively of discrete components.

2M—Miniature/Microminiature repair program.

MONOLITHIC IC—ICs that are formed completely within a semiconductor substrate. Silicon chips.

OFF-LINE TEST EQUIPMENT—Tests andisolates faults in modules or assemblies removed from systems.

OHMS PER SQUARE—The resistance of any square area of thin film resistive material as measured between two parallel sides.

ON-LINE TEST EQUIPMENT—Continuously monitors the performance of electronic systems.

ORGANIZATIONAL-LEVEL MAINTENANCE (SM&R Code O)—Responsibility of the user organization.

PACKAGING LEVELS—System developed to assist maintenance personnel in isolating faults.

PHOTO ETCHINGS—Chemical process of removing unwanted material in producing printed circuit boards.

POINT-TO-POINT WIRING—Individual wires run from terminal to terminal to complete a circuit.

PRINTED CIRCUIT BOARD (peb)—The general term for completely processed printed circuit or printed wiring configurations. It includes single-layered, double-layered, and multi-layered boards.

SCREENING—Process of applying nonconductive or semiconductive materials to a substrate to form thick film components.

SHIELDING—Technique designed to minimize internal and external interference.

SOURCE, MAINTENANCE, AND RECOVER-ABILITY CODES (SM&R CODES)—Specify maintenance level for repair of components or assemblies.

SUBSTRATE—Mounting surface for integrated circuits. May be semiconductor or insulator material depending on type of IC.

THICK FILM COMPONENTS—Passive circuit components (resistors and capacitors) having a thickness of 0.001 centimeter.

THIN FILM COMPONENTS—Passive circuit elements (resistors and capacitors) deposited on a substrate to a thickness of 0.0001 centimeter.

VACUUM EVAPORATION—Process of producing thin film components.

VERY LARGE SCALE INTEGRATION (vlsi)—An integrated circuit containing over 2,000 logic gates or 64,000 bits of memory.

WAFER—A slice of semiconductor material upon which monolithic ICs are produced.

AI-4

APPENDIX II

REFERENCE LIST

CHAPTER ONE

Linear Integrated Circuits, Basic Electricity and Electronics Course, Module 34, CANTRAC A-100-0010, Naval Education and Training Program Development Center Detachment, Great Lakes, Ill., 1981.

Technical Manual, Miniature/Microminiature (2M) Electronic Repair Program, Vols. I, II, and III,

NAVSEA TE000-AA-HBK-010/020/030/2M, Naval Sea Systems Command, Keyport, Wash., 1982.

CHAPTER 2

Technical Manual, Miniature/Microminiature (2M) Electronic Repair Program, Vols. I, II, and III,

NAVSEA TE000-AA-HBK-010/020/030/2M, Naval Sea Systems Command, Keyport, Wash., 1982.

CHAPTER 3

General Maintenance Handbook, Electronics Installation and Maintenance Books, NAVSEA SE000-00-EIM-160, Naval Sea Systems Command, Washington, D.C., 1981.

Technical Manual, Miniature/Microminiature (2M) Electronic Repair Program, Vols. I, II, and III,

NAVSEA TE000-AA-HBK-010/020/030/2M, Naval Sea Systems Command, Keyport, Wash., 1982.

AIM

Assignment Questions

Information : The text pages that you are to study are provided at the beginning of the assignment questions.

ASSIGNMENT 1

Textbook assignment: Chapter 1, "Microelectronics," pages 1-1 through 1-56. Chapter 2, "Miniature/Microminiature (2M) Repair Program and High-Reliability Soldering," pages 2-1 through 2-22.

1 -1. What term is used to describe electronic systems that are made up of extremely small parts or elements?

1. Microelectronics

2. Modular packages

3. Integrated circuits

4. Solid-state technology

1 -2. During World War II, which of the following limitations were considered unacceptable for military electronics systems?

1. Large size, heavy weight, and wide bandwidth

2. Excessive power requirements, large size, and complex manning requirements

3. Large size, heavy weight, and excessive power requirements

4. Heavy weight, complex circuits and limited communications range

1-3. The development of which of the

following types of components had the greatest impact on the technology of microelectronics?

1. Vacuum tubes and resistors

2. Transformers and capacitors

3. Vacuum tubes and transistors

4. Transistors and solid-state diodes

1 -4. For a vacuum tube to operate properly in a variety of different circuit applications, additional components are often required to "adjust" circuit values. This is because of which of the following variations within the vacuum tube?

    1. Element size

    2. Warm-up times

    3. Plug-in mountings

    4. Output characteristics

1 -5. Point to point wiring in a vacuum tube circuit often caused which of the following unwanted conditions?

    1. Heat interactions

    2. Inductive interactions

    3. Capacitive interactions

    4. Both 2 and 3 above

1 -6. Functional blocks of a system that can easily be removed for troubleshooting and repair are called

    1. sets

    2. chassis

    3. modules

    4. vacuum tubes

1 -7. Which of the following characteristics 1-11 of a printed circuit board (pcb) is NOT an advantage over a point-to-point wired tube circuit?

    1. The pcb weighs less

    2. The pcb eliminates the need for point-to-point wiring

    3. The pcb eliminates the need for a ^.\2 heavy metal chassis

    4. The pcb contains a limited number of components

1-8. A module in which the components are supported by end plates is referred to as

    1. a pcb

    2. cordwood

    3. a substrate

    4. encapsulated

1-9. A module which is difficult to repair because it is completely imbedded in solid material is one which has been

    1. balanced

    2. enveloped 2 24

    3. integrated

    4. encapsulated

1-10. All components and interconnections are formed on or within a single substrate in which of the following units?

    1. Cordwood

    2. Integrated circuit

    3. Equivalent circuit

    4. Printed circuit board

Monolithic integrated circuits are usually referred to as

    1. hybrids

    2. substrates

3. silicon chips

4. selenium rectifiers

In integrated circuits, a conductive or nonconductive film is used for which of the following types of components?

1. Capacitors and diodes

2. Transistors and diodes

3. Resistors and capacitors

4. Resistors and transistors

Which of the following types of electronic circuits is NOT a hybrid integrated circuit?

1. Thick film and transistors

2. Thin film and silicon chips

3. Transistors and vacuum tubes

4. Silicon chips and transistors

What maximum number of logic gates should be expected in a large-scale integration circuit?

1. 20

2. 200

3. 2,000

4. 20,000

1-15. Integrated circuits containing more than 64,000 bits of memory are referred to as

1. hybrid integration

2. large-scale integration

3. small-scale integration

4. very large-scale integration

1-16. Which of the following pieces of equipment is used to prepare component layout in complex ICs?

1. A mask

2. A camera

3. A computer

4. A microscope

1-17. A device that allows the depositing of material in selected areas of a semiconductor substrate, but not in others, is known as a

1. blind

2. screen

3. filter

4. wafer mask

1-18. Which of the following types of material is preferred for film circuit substrates?

1. Silicon

2. Ceramic

3. Germanium

4. Fiberglass

1-19. A typical silicon wafer has approximately (a) what diameter and (b) what thickness?

1. (a) 2 inches (b) 0.01 to 0.02 inches
2. (a) 2 inches (b) 0.21 to 0.40 inches
3. (a) 3 inches (b) 0.21 to 0.40 inches
4. (a) 3 inches (b) 0.01 to 0.20 inches

1-20. Artificially grown silicon or germanium crystals are used to produce substrates for which of the following types of integrated circuits?

1. Hybrid
2. Thin-film
3. Thick-film
4. Monolithic

1-21. Elements penetrate the semiconductor substrate in (a) what type of IC but (b) do NOT penetrate the substrate in what type ofIC?

1. (a) Diffused (b) thin-film
2. (a) Diffused (b) epitaxial
3. (a) Thick-film (b) epitaxial
4. (a) Thick-film (b) thin-film

1-22. Pn junctions are protected from contamination during the fabrication process by which of the following materials?

1. Oxide
2. Silicon
3. Germanium
4. Photoetch

1-23. The prevention of unwanted interaction or leakage between components is accomplished by which of the following techniques?

1. Isolation
2. Insulation
3. Integration
4. Differentiation

1 -24. Vacuum evaporation and cathode sputtering are two methods used to produce which of the following types of components?

1. Diodes
2. Thin-film
3. Thick-film
4. Transistors

1-25. To deposit highly reactive materials on a substrate, which of the following methods is used?

1. Photoetching
2. Photolithography
3. Cathode sputtering
4. Vacuum evaporation

1-26. To produce thin film resistors, which of the following materials is/are used?

1. Nichrome
2. Tantalum
3. Titanium
4. Each of the above

1-27. Which of the following is a major advantage of hybrid ICs?

1. Ease of manufacture

2. Ease of replacement

3. Design flexibility

4. Easy availability

1 -28. IC packaging is required for which of the following reasons?

1. To dissipate heat

2. For ease of handling

3. To increase shelf life

4. To meet stowage requirements

IN ANSWERING QUESTIONS 1-29 AND 1-30, MATCH THE IC PACKING EXAMPLES IN THE QUESTIONS TO THE PACKAGING DESCRIPTIONS IN FIGURE 1A.

Figure 1 A. —Packaging descriptions.

1-29.

1-30.

1. A

2. B

3. C

4. D

Which of the following types of DIPs are most commonly used in the Navy's microelectronics systems?

1. Glass

2. Metal

3. Ceramic

4. Plastic

In IC production, gold or aluminum bonding wires are used for which of the following purposes?

1. To bond the chip to the package

2. To provide component isolation

3. To connect the package to the circuit board

4. To connect the chip to the package leads

1-33. IC packages that may be easily installed by hand or machine on mounting boards fall into which of the following categories?

1. TO

2. DIP

3. Flatpack

4. Each of the above

1-34. The need for bonding wires has been eliminated by which of the following production techniques?

1. LSI

2. Beam lead

3. Flip chip

4. Both 2 and 3 above

THIS SPACE LEFT BLANK INTENTIONALLY.

IN ANSWERING QUESTIONS 1-35 THROUGH 1-37, MATCH THE LETTER IN EACH OF THE FIGURES THAT IDENTIFIES PIN 1.

1-35.

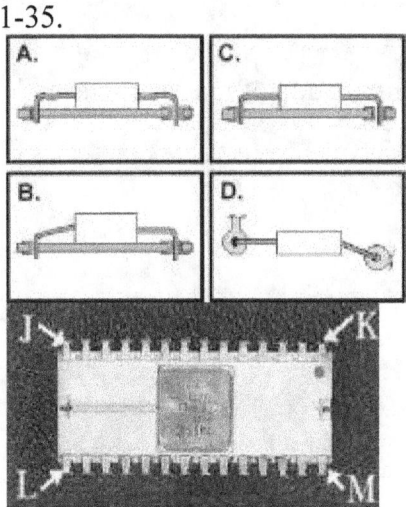

1-38.

1. 2. 3. 4.

J

K L M

Letters and numbers stamped on the body of an IC serve to provide which of the following types of information?

1. A

2. B

3. C

4. D

1. Use

2. Serial number

3. Date of manufacture

4. Applicable equipment

1-36.

1-39.

1. 2. 3. 4.

E F G H

1-40.

Descriptive information about a particular type of IC may be found in which of the following documents?

1. The manufacturer's data sheet
2. The equipment Allowance Part List (APL)
3. The National Stock Number (NSN)
4. The IC identification number list

Assemblies made up EXCLUSIVELY of discrete electronic parts are classified as

1. vacuum-tube circuits
2. microcircuit modules
3. hybrid microcircuits
4. miniature electronics circuits

6

1-41. An assembly of microcircuits or a combination of microcircuits and discrete components is referred to as a

1. mother board
2. microprocessor
3. miniature module
4. microcircuit module

1 -42. A technician has isolated a problem to a plug-in module on a printed circuit board. What is this level of system packaging?

1. Level O
2. Level I
3. Level II
4. Level III

1 -43. A faulty transistor would be identified as what level of packaging?

1. Level O
2. Level I
3. Level II
4. Level III

1 -44. A chassis located in a radar antenna pedestal would be identified as what level of system packaging?

1. Level I
2. Level II
3. Level III
4. Level IV

1-45. Which of the following characteristics is NOT an advantage of multilayer printed circuit boards?

1-46. Which of the following circuit connection methods is NOT used in making interconnections on a multilayer printed circuit board interconnection?

1. Terminal lug
2. Clearance hole
3. Layer build-up
4. Plated-through hole

1-47. The most complex to produce and
difficult to repair printed circuit boards are those made using which of the following methods?

1. Layer-buildup
2. Clearance-hole
3. Step-down-hole
4. Plated-through-hole

1-48. Environmental performance
requirements for ICs are set forth in which of the following publications?

1. 2M repair manual
2. Military Standards
3. System maintenance manuals
4. Manufacturer's data sheet

1 -49. Ground planes and shielding are used to prevent which of the following electrical interactions?

1. Cross talk
2. External interference
3. The generation of rf within the system
4. All of the above

1-50. Training requirements for miniature and microminiature (2M) repair personnel was established by which of the following authorities?

1. Chief of Naval Education and Training
2. Chief of Naval Technical Training
3. Chief of Naval Operations
4. Commander, Naval Sea Systems Command

1-51. The standards of workmanship and guidelines for specific repairs to equipment are contained in which of the following Navy publications?

1. Introduction to Microelectronics
2. NAVSHIPS Technical Manual
3. Electronics Installation and Maintenance Books (EIMB)
4. Miniature/Microminiature (2M) Electronic Repair Program

1-52. A technician is authorized to perform
2M repairs upon satisfactory completion of which of the following types of training?

1. A 2M training class
2. On-the-job training
3. NEETS, Module 14
4. Any electronics class "A" school

1-53. Repairs that are limited to discrete components and single- and double-sided boards are classified as what level of repairs?

1. Intermediate
2. Organizational
3. Miniature component
4. Microminiature component

1-54. To ensure that a 2M technician
maintains the minimum standards of workmanship, the Navy requires that the

technician meet which of the following requirements?

1. Be licensed
2. Be certified
3. Be experienced
4. Be retrained

1-55. If a technician should fail to maintain the required standards of workmanship, the technician's certification is subject to what action?

1. Cancellation
2. Recertification
3. Reduction to next lower level
4. Withholding pending requalification

1-56. The most extensive shop facilities and highly skilled technicians are located at what SM & R level of maintenance?

1. Depot
2. Operational
3. Intermediate
4. Organizational

1-57. SM & R code D maintenance facilities are usually located at which of the following activities?

1. Shipyards
2. Contractor maintenance organizations
3. Shore-based facilities
4. All of the above

1-58. Direct support to user organizations is provided by which of the following SM & R code maintenance levels?

1. Depot
2. Operational
3. Intermediate
4. Organizational

1-59. Inspecting, servicing, and adjusting equipment is the function of which of the following SM & R code maintenance levels?

1. Depot
2. Operational
3. Intermediate
4. Organizational

1-60. The maintenance level at which normal 2M repairs are performed is set forth in the maintenance plan and specified by the

1. NAVSEA 2M Repair Program
2. Source, Maintenance, and Recoverability (SM & R) code
3. Chief of Naval Operations
4. Equipment manufacturers' documentation

1-61. Boards or modules that are SM & R code D may be repaired at the organizational level under which of the following conditions?

1. On a routine basis
2. When parts are available

3. To meet an urgent operational commitment

4. When code D repair will take six weeks or longer

1-62. Source, Maintenance, and Recovery (SM & R) codes that list where repair parts may be obtained, who is authorized to make the repair, and the maintenance level for the item are found in which of the following documents?

1. Allowance Equipage Lists (AEL)

2. Allowance Parts Lists (APL)

3. Manufacturer's Parts List

4. Navy Stock System

1-63. Test equipment that continuously monitors performance and automatically isolates faults to removable assemblies is what category of equipment?

1. On-line

2. Off-line

3. General purpose

4. Fault isolating

1-64. A dc voltmeter that is permanently attached to a power supply for the purpose of monitoring the output is an example of what type test equipment?

1. General Purpose Electronic Test Equipment (GPETE)

2. Built in Test Equipment (BITE)

3. Off-line test equipment

4. Specialized test equipment

1-65. Which of the following types of test equipment is classified as off-line automatic test equipment?

1. Centralized Automatic Test System (CATS)

2. Versatile Avionic Shop Test System (VAST)

3. General Purpose Electronic Test Equipment (GPETE)

4. Test Evaluation and Monitoring System (TEAMS)

1-66. Fault diagnosis using GPETE should only be attempted by which of the following personnel?

1. Officers

2. Technician strikers

3. Experienced technicians

4. Basic Electricity and Electronics school graduates

1-67. During fault isolation procedures, a device or component should be desoldered and removed from the circuit only at which of the following times?

1. After defect verification

2. For out-of-circuit testing

3. During static resistance checks

4. At any time the technician desires

1-68. 2M repair stations are equipped according to the types of repairs to be accomplished. The use of microscopes and precisions drill presses would be required in which of the following types of repair?

1. Miniature

2. Microminiature

3. Both 1 and 2 above

4. Emergency

1-69. In the selection of a soldering iron tip, which of the following factors should be considered?

1. The complexity of the pcb
2. The composition of the pcb
3. The area and mass being soldered
4. The type of component being soldered

1-70. The handpiece that can be used for the greatest variety of operations is the

1. solder extractor
2. rotary-drive tool
3. resistive tweezers
4. lap flow and thermal scraper handtool

1-71. Regardless of location, 2M repair stations require adequate work surface area, lighting, power, and what other minimum requirement?

1. Heat source
2. Ventilation
3. Illumination
4. Dust-free space

1-72. Solder used in electronics is an alloy composed of which of the following metals?

1. Tin and zinc
2. Tin and lead
3. Lead and zinc
4. Lead and copper

1-73. A roll of solder is marked 60/40. What do these numbers indicate?

1. 60% tin, 40% lead
2. 60% tin, 40% copper
3. 60% lead, 40% tin
4. 60% lead, 40% copper

1-74. Which of the following alloys will melt directly into a liquid and have no plastic or semiliquid state?

1. Metallic alloy
2. Eutectic alloy
3. Zinc-lead alloy
4. Copper-zinc alloy

1-75. The PREFERRED solder alloy ratio for electronic repair is 63/37. Which of the following alloy ratios is also ACCEPTABLE for this type of repair?

1. 30/70
2. 50/50
3. 60/40
4. 70/30

ASSIGNMENT 2

Textbook assignment: Chapter 3, "Miniature and Microminiature Repair Procedures," pages 3-1 through 3-51.

2-1. Which of the following requirements must be met by a 2M technician to be authorized to perform repairs at a particular level?

1. Have knowledge of that level
2. Be certified at that level
3. Be experienced at that level
4. Be licensed at the next higher level

2-2. Protective materials applied to electronic assemblies to prevent damage caused by corrosion, moisture, and stress are called

1. conformal coatings
2. isolation materials
3. electrical insulation
4. encapsulation coatings

2-3. Before working on a pcb, the conformal coating should be removed from what part of the board?

1. The entire board
2. The component side
3. The area of the repair
4. The side opposite the component side

2-4. What are the approved methods of conformal coating removal?

1. Peeling and abrading
2. Stripping and heating
3. Mechanical and thermal only
4. Mechanical, thermal, and chemical

2-5. Most methods of conformal coating removal are variations of which of the following types of removal?

1. Thermal
2. Chemical
3. Mechanical
4. Electrical

2-6. What is the preferred method of removing epoxy conformal coatings?

1. Solvent
2. Ball mill
3. Hot-air jet
4. Thermal parting tool

2-7. A conformal coating is considered to be thin if it is less than what thickness?

1. 0.025 inches
2. 0.040 inches
3. 0.050 inches
4. 0.250 inches

2-8. Of the various mechanical methods of conformal coating removal, physical contact with the work piece is NOT required using which of the following methods?

1. Cutting
2. Grinding
3. Hot-air jet
4. Thermal parting

2-9. Cutting and peeling is an easy method of removing which of the following types of coatings?

1. Epox
2. Varnish
3. Parylene
4. Silicone

2-10. Thin acrylic coatings are readily removed 2-15 in which of the following ways?

1. Routing
2. Hot-air jet
3. Cut and peel
4. Mild solvents

2-11. When applying an application of conformal coating, which of the following 2-16 conditions is true?

1. The board should be clean and moist
2. The coating should be applied only to the component replaced
3. The coating should be the same type as that used by the manufacturer
4. The coating should be applied only to the solder joints 2-17

2-12. The procedure of connecting circuitry on one side of a board with the circuitry on the other side is known as

1. mounting
2. termination
3. hole reinforcement
4. interfacial connections 2-18

2-13. Reinforcement for circuit pads on both sides of the board is provided by which of the following types of eyelets?

1. Flat-set
2. Roll-set
3. Funnel-set

2-19

2-14. The manner in which wires and leads are attached to an assembly is described by which of the following terms?

1. Termination
2. Solderjoints
3. Lead formation
4. Interfacial connections

Clinched leads, straight-through leads, and offset pads are variations of what type of termination?

1. Solder cup
2. On-the-board
3. Above-the-board
4. Through-hole

What total number of degrees of bend are (a) fully clinched leads and (b) semiclinched leads?

Which of the following types of terminals are used as tie points for interconnecting wiring?

1. Pins
2. Hooks
3. Solder cups
4. Turrets

During assembly, component stability is provided by which of the following types of lead termination?

1. Hook terminal
2. Clinched lead
3. Turret terminal
4. On-the-board (lap-flow)

Which of the following types of lead terminations is the easiest to remove and rework?

1. Offset-pad
2. Semiclinched
3. Fully clinched
4. Straight-through

2-20. Turret, fork, and hook terminals are 2-25 examples of what type of termination?

1. Off-set
2. On-the-board
3. Above-the-board
4. Through-the-board

2-21. When a lead is soldered to a pad without passing it through the board, what type of termination has been made? 2-26

1. Lap-flow
2. Off-set pad
3. Clinched lead
4. Straight-through lead

2-22. Most damage to printed circuit boards occurs at which of the following times?

2-27

1. During trouble shooting
2. During component removal
3. During component replacement
4. During normal system operations

2-23. Of the solder removal methods listed below, which one is most versatile and reliable?

2-28

1. Wicking
2. Heat and shake
3. Motorized solder extractor
4. Manually controlled vacuum plunge

2-24. When removing solder with a solder wick, where should the wick be placed in relation to the solder joint and the iron?

1. Below both the j oint and the iron
2. Between the joint and the iron

3. Above the joint and the iron

The motorized solder extractor may be operated in three different modes. Which of the following is NOT one of those modes?

1. Hot-air jet
2. One-shot vacuum
3. Heat and vacuum
4. Heat and pressure

Stirring the lead during desoldering prevents which of the following unwanted results?

1. Sweat joints
2. Overheating
3. Cold solder joints
4. Pad delamination

Of the following solder removal methods, which one is acceptable for removing solder?

1. Hot-air jet
2. Heat and pull
3. Heat and shake
4. Heat and squeeze

Which of the following examples represents properly formed leads?

3.

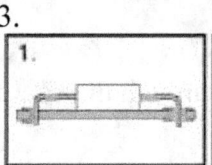

2-29. A 2M technician is repairing a board that is manufactured with semiclinched leads. What type of termination should the technician use in replacing components?

1. Lap flow
2. Clinched lead
3. Semiclinched lead
4. Straight-through-lead

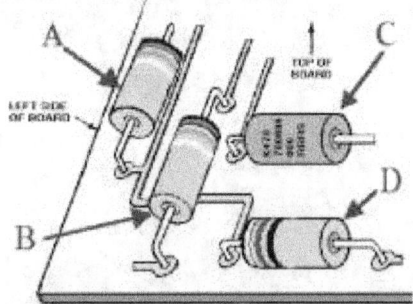

Figure 2A.—Component mounting.

IN ANSWERING QUESTION 2-30, REFER TO FIGURE 2A.

2-30. What component on the board is NOT properly mounted?

1. A
2. B
3. C

4. D

2-31. What type of heat source is a soldering iron?

1. Radiant
2. Resistive
3. Conductive
4. Convective

2-32. The shape and size of the soldering iron tip to be used is determined by which of the following factors?

1. The type of flux to be used
2. The type of work to be done
3. The type of solder to be used
4. The voltage source for the iron

2-33. A thermal shunt is attached to the leads of a transistor prior to applying solder to the joint. This shunt serves what purpose?

1. Prevents short circuits
2. Retains heat at the joint
3. Conducts heat away from the component
4. Physically supports the lead during soldering

2-34. What is the appearance of a good solder joint?

1. Dull gray
2. Crystalline
3. Bright and shiny
4. Shiny with small pits

2-35. What is the most efficient soldering temperature?

1. 360 °F
2. 440 °F
3. 550 °F
4. 800 °F

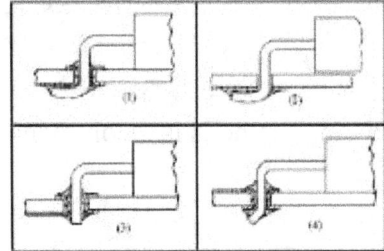

Figure 2B. —Solder joints.

IN ANSWERING QUESTION 2-36, REFER TO FIGURE 2B.

2-36. All the solder joints in the figure have two things in common. The first is that all are through-the-board terminations. The other is that they are all what type of joint?

1. Sweat
2. Full-fillet
3. Unacceptable
4. Clinched-lead

2-37. Plug-in DIPs are mounted by using which of the following aids/parts?

1. Insulators
2. Special tools

3. Clinched leads

4. Mounting sockets

2-38. Plug-in DIPs are susceptible to loosening because of which of the following causes?

1. Heat

2. Stress

3. Warpage

4. Vibration

2-39. Component leads may be clipped to aid in their removal under which of the following conditions?

1. When the component is conformally coated

2. When the component is known to be defective

3. When board damage may result from normal removal methods

4. Both 2 and 3 above

2-40. Visual inspection of a completed repair is conducted to evaluate which of the following aspects of the repair?

1. Workmanship quality

2. Component placement

3. In-circuit quality test

4. Conformal coating integrity

2-41. When speaking of TO mounting techniques, the term "plug-in" refers to the same technique as is used with DIPs.

1. True

2. False

2-42. In addition to heat dissipation and physical support, which of the following needs might justify the use of a spacer with a TO mount?

1. Vibration elimination

2. Proper lead formation

3. Proper lead termination

4. Short-circuit protection

2-43. For the removal of imbedded TOs in which all the leads are free, which of the following methods is recommended?

1. Push out gently

2. Pull out with pliers

3. Pull out with fingers

4. Tap out with soft mallet

2-44. What is the most critical step in replacing an imbedded TO?

1. Seating the leads

2. Forming the leads

3. Soldering the leads

4. Seating the body of the TO

2-45. When a TO or a DIP is replaced on a printed circuit board, what type of termination is normally used?

1. Lap flow

2. On-the-board

3. Above-the-board

4. Through-the-board

2-46. What type of termination is used in the replacement of a flat pack?

1. Lap flow
2. Solder cup
3. Solder plug
4. Full fillet

2-47. Heating the leads and lifting them free with tweezers is the preferred method of removing which of the following components?

1. TOs
2. DIPs
3. Flat packs
4. Transistors

2-48. Which of the examples shows the correct lead formation for a flat pack?

2-49. Use of a skipping pattern when soldering multilead components prevents

1. cold solder joints
2. excessive heat buildup
3. the component from moving
4. the need to visually inspect the piece

2-50. Cards and boards may be damaged under which of the following conditions?

1. When unauthorized repairs are attempted by untrained personnel
2. When technicians use improper tools
3. When improperly stored
4. Each of the above

2-51. DS3 Spark is preparing to repair a cracked conductor on a card. For this type of damage, which of the following repair methods is preferred?

1. Solder bridge
2. Clinched staple
3. Install an eyelet at the crack and solder in place
4. Lap-flow soldered wire across the break

2-52. To ensure a good mechanical bond between the board and replacement pad and to provide good electrical contact for components, which of the following procedures is used?

1. Epoxying the pad to the run
2. Electroplating the repair area
3. Installing an eyelet in the pad
4. Lap-flow soldering the repair area

2-53. Breaks, holes, and cracks in pcbs are repaired by using a mixture of

1. fiberglass and rosin
2. epoxy and powdered carbon
3. conformal coating and RTV
4. epoxy and powdered fiberglass

2-54. What is the first step in the repair of burned or scorched boards?

1. Filling the burned area with epoxy and fiberglass
2. Removing all discolored material
3. Removing all delaminated conductors

4. Cleaning all surfaces with solvent

2-57.

Electrostatic discharge (ESD) has the greatest effect on which of the following devices?

1. Silicon diodes
2. Selenium rectifiers
3. Germanium transistors
4. Metal-oxide transistor

Figure 2C. —Symbol.

IN ANSWERING QUESTION 2-58, REFER TO FIGURE 2C.

2-55.

2-56.

Which of the following statements is correct concerning the repair of repairable delaminated conductors?

1. 2.

4.

All delaminations are removed Repairable delaminations are not removed

All delaminations are epoxied to the board

All delaminations are replaced with insulated wire

Damage to some electronic components can occur at what minimum electrostatic potential?

1. 14 volts
2. 35 volts
3. 350 volts
4. 3,500 volts

2-58. The symbol shown in the figure is found on a component package. What does it indicate about the component?

1. It is a high cost item
2. It is a bipolar device
3. It is an integrated circuit
4. It is electrostatic discharge sensitive

2-59. Electrostatic charges may develop as high as which of the following voltages?

1. 3,000 volts
2. 15,000 volts
3. 20,000 volts
4. 35,000 volts

2-60. To prevent an electrostatic charge built up on the body of the technician from damaging ESDS devices, the technician should take which of the following precautions?

    1. Be grounded

    2. Wear gloves

    3. Ground the device

    4. Handle the device with insulated tools

    2-61. The best source of information concerning the application, handling, and storage of any aerosol dispensers may be found in/on which of the following sources?

    1. NAVSEA 2M manual

    2. NEETS, Module 14, Topic 3

    3. On the aerosol dispenser

    4. Applicable military standards

www.ingramcontent.com/pod-product-compliance
Lightning Source LLC
Chambersburg PA
CBHW081607200526
45169CB00021B/2212